THE NAMES OF PLANTS

The Names of Plants is a handy, two-part reference book for the botanist and amateur gardener. The book begins by documenting the historical problems associated with an ever-increasing number of common names of plants and the resolution of these problems through the introduction of International Codes for both botanical and horticultural nomenclature. It also outlines the rules to be followed when plant breeders name a new species or cultivar of plant.

The second part of the book comprises an alphabetical glossary of generic and specific plant names, and components of these, from which the reader may interpret the existing names of plants and construct new names.

For the third edition, the book has been updated to include explanations of the International Codes for both Botanical Nomenclature (2000) and Nomenclature for Cultivated Plants (1995). The glossary has similarly been expanded to incorporate many more commemorative names.

THE NAMES OF PLANTS

THIRD EDITION

David Gledhill
Formerly
Senior Lecturer, Department of Botany, University of Bristol and
Curator of Bristol University Botanic Garden

PUBLISHED BY THE PRESS SYNDICATE OF THE UNIVERSITY OF CAMBRIDGE
The Pitt Building, Trumpington Street, Cambridge, United Kingdom

CAMBRIDGE UNIVERSITY PRESS
The Edinburgh Building, Cambridge CB2 2RU, UK
40 West 20th Street, New York, NY 10011-4211, USA
477 Williamstown Road, Port Melbourne, VIC 3207, Australia
Ruiz de Alarcón 13, 28014 Madrid, Spain
Dock House, The Waterfront, Cape Town 8001, South Africa

http://www.cambridge.org

First published 1985
Second edition 1989
Reprinted 1990, 1992, 1994, 1996
Third edition 2002

Printed in the United Kingdom at the University Press, Cambridge

Typeface Joanna MT 10/13 pt   *System* LaTeX 2ε [TB]

A catalogue record for this book is available from the British Library

Library of Congress Cataloguing in Publication data

Gledhill, D.
The names of plants / David Gledhill. – 3rd ed.
p. cm.
Includes bibliographical references (p.) and index.
ISBN 0 521 81863 X ISBN 0 521 52340 0 (pbk.)
1. Botany – Nomenclature. 2. Botany – Dictionaries – Latin (Medieval and modern)
3. Latin language, Medieval and modern – Dictionaries – English. I. Title.
QK96 .G54 2002 580′.14 – dc21 2002022290

ISBN 0 521 81863 X hardback
ISBN 0 521 52340 0 paperback

Contents

Preface to the first edition

Originally entitled *The naming of plants and the meanings of plant names*, this book is in two parts. The first part has been written as an account of the way in which the naming of plants has changed with time and why the changes were necessary. It has not been the writer's intention to dwell upon the more fascinating aspects of common names but rather to progress from these to the situation which exists today; in which the botanical and horticultural names of plants must conform to internationally agreed standards. The aim has been to produce an interesting text which is equally as acceptable to the amateur gardener as to the botanist. The temptation to make this a definitive guide to the International Code of Botanical Nomenclature was resisted since others have done this already and with great clarity. A brief comment on synonymous and illegitimate botanical names and a reference to recent attempts to accommodate the various traits and interests in the naming of cultivated plants was added after the first edition.

The book had its origins in a collection of Latin plant names, and their meanings in English, which continued to grow by the year but which could never be complete. Not all plant names have meaningful translations. Some of the botanical literature gives full citation of plant names (and translations of the names, as well as common names). There are, however, many horticultural and botanical publications in which plant names are used in a casual manner, or are misspelled, or are given meanings or common names that are neither translations nor common (in the world-wide sense). There is also a tendency that may be part of modern language, to reduce names of garden plants to an abbreviated form (e.g. Rhodo for *Rhododendron*). Literal names such as Vogel's *Napoleona*, for *Napoleona vogelii*, provide only limited information about the plant. The dedication of the genus to Napoleon Bonaparte is not informative. Only by further search of the literature will the reader find that Theodor Vogel was the botanist to the 1841 Niger expedition and that he collected some 150 specimens during a rainy July fortnight in Liberia. One of those specimens, number 45, was a *Napoleona* that was later named for him as the type of the new species by Hooker and Planchon. To have given such information would have made the text very much larger.

The author has compiled a glossary which should serve to translate the more meaningful and descriptive names of plants from anywhere on earth but which will give little information about many of the people and places commemorated in plant names. Their entries do little more than identify the persons for whom the names were raised and their period in history. The author makes no claim that the glossary is all-encompassing or that the meanings he has listed are always the only meanings that have been put upon the various entries. Authors of Latin names have not always explained the meanings of the names they have erected and, consequently, such names may have been given different meanings by subsequent writers.

Preface to the third edition

Since making the assumption, in the second edition, that genetic manipulation of the properties of plants might require new consideration of the ways in which they are to be named, GM has proceeded apace. Not only can the innate genetic material be re-ordered – in ways that nature would have rejected through their exposure to natural selection by the environment – but alien genetic material, from other organisms, can be introduced to give bizarre results. *Arabidopsis thaliana* has only 10 chromosomes and has been the plant of choice for cytologists and nucleic acid workers because of this. The twenty-first century sees its genetic code mapped and its 25,000 genes being examined individually to ascertain the 'meaning of plant life'. From quite practical beginnings such as giving tomato fruits an extended keeping time, to esoteric developments such as building a luminescence gene from a jellyfish into a mouse, there is now a proposal to insert a gene from an electric eel into plants so that the plants can provide sources of electricity. This new 'green revolution' has an historical ring of familiarity about it!

The new century has not yet brought universal consistency in accepting the botanical and the horticultural codes. Yet science is already seeking to move towards an international biodiversity code for the naming of everything. If one was to be facetious, one might observe that man is still at 6's and 7's in seeking an explanation of everything – and may well, in the end, find that the answer is 42!

The study of whole organisms and their systematic relationships is an economically unrewarding pure science but an essential area of continuing investigation. If man is intent on producing genetically deviant life forms, the descent of these must be known and their names must reflect that descent.

The nature of the problem

A rose: by any name?

Man's highly developed constructive curiosity and his capacity for communication are two of the attributes distinguishing him from all other animals. Man alone has sought to understand the whole living world and things beyond his own environment and to pass his knowledge on to others. Consequently, when he discovers or invents something new he also creates a new word, or words, in order to be able to communicate his discovery or invention to others. There are no rules to govern the manner in which such new words are formed other than those of their acceptance and acceptability. This is equally true of the common, or vulgar or vernacular names of plants. Such names present few problems until communication becomes multilingual and the number of plants named becomes excessive. For example, the diuretic dandelion is easily accommodated in European languages. As the lion's tooth, it becomes Lowenzahn, dent de lion, dente di leone. As piss-abed it becomes Pissenlit, piscacane, and piscialetto. When further study reveals that there are more than a thousand different kinds of dandelion throughout Europe, the formulation of common names for these is both difficult and unacceptable.

Common plant names present language at its richest and most imaginative (welcome home husband however drunk you be, for the houseleek or *Sempervivum*; shepherd's weather-glass, for scarlet pimpernel or *Anagallis*; meet her i'th'entry kiss her i'th'buttery, or leap up and kiss me, for *Viola tricolor*; touch me not, for the balsam *Impatiens noli-tangere*; mind your own business, or mother of thousands, for *Soleirolia soleirolii*; blood drop emlets, for *Mimulus luteus*). Local variations in common names are numerous and this is perhaps a reflection of the importance of plants in general conversation, in the kitchen and in herbalism throughout the country in bygone days. An often quoted example of the multiplicity of vernacular names is that of *Caltha palustris*, for which, in addition to marsh marigold, kingcup and May blobs, there are 90 other local British names (one being dandelion), as well as over 140 German and 60 French vernacular names.

Common plant names have many sources. Some came from antiquity by word of mouth as part of language itself, and the passage of time and changing circumstances have obscured their meanings. Fanciful ideas of a plant's association with animals, ailments and festivities, and observations of plant structures, perfumes, colours, habitats and seasonality have all contributed to their naming. So too have their names in other languages. English plant names have come from Arabic, Persian, Greek, Latin, ancient British, Anglo-Saxon, Norman, Low German, Swedish and Danish. Such names were introduced together with the spices, grains, fruit plants and others which merchants and warring nations introduced to new areas. Foreign names often remained little altered but some were transliterated in such a way as to lose any meaning which they may have had originally.

The element of fanciful association in vernacular plant names often drew upon comparisons with parts of the body and with bodily functions (priest's pintle for *Arum maculatum*, open arse for *Mespilus germanicus* and arse smart for *Polygonum hydropiper*). Some of these persist but no longer strike us as 'vulgar' because they are 'respectably' modified or the associations themselves are no longer familiar to us (*Arum maculatum* is still known as cuckoo pint (cuckoo pintle) and as wake robin). Such was the sensitivity to indelicate names that Britten and Holland, in their *Dictionary of English Plant Names* (1886), wrote 'We have also purposely excluded a few names which though graphic in their construction and meaning, interesting in their antiquity, and even yet in use in certain counties, are scarcely suited for publication in a work intended for general readers'. They nevertheless included the examples above. The cleaning up of such names was a feature of the Victorian period, during which our common plant names were formalized and reduced in numbers. Some of the resulting names are prissy (bloody cranesbill, for *Geranium sanguineum*, becomes blood-red cranesbill), some are uninspired (naked ladies or meadow saffron, for *Colchicum autumnale*, becomes autumn crocus) and most are not very informative.

This last point is not of any real importance because names do not need to have a meaning or be interpretable. Primarily, names are mere ciphers which are easier to use than lengthy descriptions and yet, when accepted, they can become quite as meaningful. Within limits, it is possible to use one name for a number of different things but, if the limits are exceeded, this may cause great confusion. There are many common plant names which refer to several plants but cause no problem so long as they are used only within their local areas or when they are used to convey only a general idea of the plant's identity. For example, *Wahlenbergia saxicola* in New Zealand, *Phacelia whitlavia* in southern California, USA, *Clitoria ternatea* in West Africa, *Campanula rotundifolia* in Scotland and *Endymion non-scriptus* (formerly *Scilla non-scripta* and now *Hyacinthoides non-scripta*) in England are all commonly called bluebells. In each area, local people will understand others who speak of bluebells but in all the areas except Scotland the song 'The Bluebells of Scotland', heard perhaps on the radio, will conjure up a wrong impression. At least ten different plants are given the common name of cuckoo-flower in England, signifying only that they flower in spring at a time when the cuckoo is first heard.

The problem of plant names and of plant naming is that common names need not be formed according to any rule and can change as language, or the user of language, dictates. If our awareness extended only to some thousands of 'kinds' of plants we could manage by giving them numbers but, as our awareness extends, more 'kinds' are recognized and for most purposes we find a need to organize our thoughts about them by giving them names and by forming them into named groups. Then we have to agree with others about the names and the groups, otherwise communication becomes hampered by ambiguity. A completely coded numerical system could be devised but would have little use to the non-specialist, without access to the details of encoding.

Formalized names provide a partial solution to the two opposed problems presented by vernacular names: multiple naming of a single plant and multiple application of a single name. The predominantly two-word structure of such formal names has been adopted in recent historic times in all biological nomenclature, especially in the branch which, thanks to Isidorus Hispalensis (560–636), Archbishop of Seville, whose 'Etymologies' was a vast encyclopaedia of ancient learning and was studied for 900 years, we now call botany. Of necessity, botanical names have been formulated from former common names but this does not mean that in the translation of botanical names we may expect to find meaningful names in common language. Botanical names, however, do represent a stable system of nomenclature which is usable by people of all nationalities and has relevancy to a system of classification.

Since man became wise, he has domesticated both plants and animals and, for at least the past 300 years, has bred and selected an ever growing number of 'breeds', 'lines' or 'races' of these. He has also given them names. In this, man has accelerated the processes which, we think, are the processes of natural evolution and has created a different level of artificially sustained, domesticated organisms. The names given by the breeders of the plants of the garden and the crops of agriculture and arboriculture present the same problems as those of vernacular and botanical names. Since the second edition of this book was published, genetic manipulation of the properties of plants has proceeded apace. Not only has the innate genetic material of plants been re-ordered, but alien genetic material, from other organisms, even from other kingdoms, has been introduced to give bizarre results. The products are unnatural and have not faced selection in nature. Indeed, some may present problems should they interbreed with natural populations in the future. There is still a divide between the international bodies concerned with botanical and cultivated plant names and the commercial interests that are protected by legislation for trademarking new genetic and transgenic products.

The size of the problem

'Man by his nature desires to know' (Aristotle)

Three centuries before Christ, Aristotle of Stagira, disciple of Plato, wrote extensively and systematically of all that was then known of the physical and living world. In this monumental task, he laid the foundations of inductive reasoning. When he died, he left his writings and his teaching garden to one of his pupils, Theophrastus (*c.* 370–285 BC), who also took over Aristotle's peripatetic school. Theophrastus' writings on mineralogy and plants totalled 227 treatises, of which nine books of *Historia Plantarum* contain a collection of contemporary knowledge about plants and eight of *De Causis Plantarum* are a collection of his own critical observations, a departure from earlier philosophical approaches, and rightly entitle him to be regarded as the father of botany. These works were subsequently translated into Syrian, to Arabic, to Latin and back to Greek. He recognized the distinctions between monocotyledons and dicotyledons, superior and inferior ovaries in flowers, the necessity for pollination and the sexuality of plants but, although he used names for plants of beauty, use or oddity, he did not try to name everything.

To the ancients, as to the people of earlier civilizations of Persia and China, plants were distinguished on the basis of their culinary, medicinal and decorative uses – as well as their supposed supernatural properties. For this reason, plants were given a name as well as a description. Theophrastus wrote of some 500 'kinds' of plant which, considering that material had been brought back from Alexander the Great's campaigns throughout Persia, as far as India, would indicate a considerable lack of discrimination. In Britain, we now recognize more than that number of different 'kinds' of moss.

Four centuries later, about AD 64, Dioscorides recorded 600 'kinds' of plants and, half a century later still, the elder Pliny, in his huge compilation of the information contained in the writings of 473 authors, described about a thousand 'kinds'. During the 'Dark Ages', despite the remarkable achievements of such people as Albertus Magnus (1193–1280), who collected plants during extensive journeys in Europe, and the publication of the *German Herbarius* in 1485 by another collector of European plants, Dr Johann von Cube, little progress was made in the study of plants. It was the renewal of critical observation by Renaissance botanists such as Dodoens (1517–1585), l'Obel (1538–1616), l'Ecluse (1526–1609) and others which resulted in the recognition of some 4,000 'kinds' of plants by the sixteenth century. At this point in history, the renewal of critical study and the beginning of plant collection throughout the known world produced a requirement for a rational system of grouping plants. Up to the sixteenth century, three factors had hindered such classification. The first of these was that the main interested parties were the nobility and apothecaries who conferred on plants great monetary value, either because of their rarity or because of the real or imaginary virtues attributed to them, and regarded them as items to be guarded jealously. Second was the lack of any standardized system of naming plants and third, and perhaps most important, any expression of the idea that living things could have evolved from earlier extinct ancestors and could therefore form groupings of related 'kinds' was a direct contradiction of the religious dogma of Divine Creation.

Perhaps the greatest disservice to progress was that caused by the Doctrine of Signatures, which claimed that God had given to each 'kind' of plant some feature which could indicate the uses to which man could put the plant. Thus, plants with kidney-shaped leaves could be used for treating kidney complaints and were grouped together on this basis. Theophrastus Bombast von Hohenheim (1493–1541) had invented properties for many plants under this doctrine. He also considered that man possessed intuitive knowledge of which plants could serve him, and how. He is better known under the Latin name which he assumed, Paracelsus, and the doctrinal book *Dispensatory* is usually attributed to him. The doctrine was also supported by Giambattista Della Porta (1543–1615), who made an interesting extension to it, that the distribution of different 'kinds' of plants had a direct bearing upon the distribution of different kinds of ailment which man suffered in different areas. On this basis, the preference of willows for wet habitats is ordained by God because men who live in wet areas are prone to suffer from rheumatism and, since the bark of *Salix* species gives relief from rheumatic pains (it contains salicylic acid, the analgesic principal of aspirin), the willows are there to serve the needs of man.

In spite of disadvantageous attitudes, renewed critical interest in plants during the sixteenth century led to more discriminating views as to the nature of 'kinds', to searches for new plants from different areas and concern over the problems of naming plants. John Parkinson (1569–1629), a London apothecary, wrote a horticultural landmark with the punning title *Paradisi in Sole – Paradisus Terrestris* of 1629. This was an encyclopaedia of gardening and of plants then in cultivation and contains a lament by Parkinson that, in their many catalogues, nurserymen 'without consideration of kind or form, or other special note give(th) names so diversely one from the other, that . . . very few can tell what they mean'. This attitude towards common names is still with us but not in so violent a guise as that shown by an unknown author who, in *Science Gossip* of 1868, wrote that vulgar names of plants presented 'a complete language of meaningless nonsense, almost impossible to retain and certainly worse than useless when remembered – a vast vocabulary of names, many of which signify that which is false, and most of which mean nothing at all'.

Names continued to be formed as phrase-names constructed with a starting noun (which was later to become the generic name) followed by a description. So, we find that the creeping buttercup was known by many names, of which Caspar Bauhin (1550–1624) and Christian Mentzel (1622–1701) listed the following:

Caspar Bauhin, *Pinax Theatri Botanici*, 1623:

Ranunculus pratensis repens hirsutus var.   C.Bauhin
repens fl. luteo simpl.   J.Bauhin
repens fol. ex albo variis
repens magnus hirsutus fl. pleno
repens flore pleno
pratensis repens Parkinson
pratensis reptante cauliculo   l'Obel
polyanthemos 1   Dodoens
hortensis 1   Dodoens
vinealis   Tabernamontana
pratensis etiamque hortensis   Gerard

Christianus Mentzelius, *Index Nominum Plantarum Multilinguis (Universalis)*, 1682:

Ranunculus pratensis et arvensis   C.Bauhin
rectus acris var.   C.Bauhin
rectus fl. simpl. luteo   J.Bauhin
rectus fol. pallidioribus hirsutis   J.Bauhin
albus fl. simpl. et denso   J.Bauhin
pratensis erectus dulcis   C.Bauhin
Ranoncole dolce   Italian
Grenoillette dorée o doux   Gallic
Sewite Woode Crawe foet   English
Suss Hanenfuss
Jaskien sodky   Polish
Chrysanth. simplex   Fuchs
Ranunculus pratensis repens hirsutus var. c   C.Bauhin
repens fl. luteo simpl.   J.Bauhin
repens fol. ex albo variis   Antonius Vallot
repens magnus hirsut. fl. pleno   J.B.Tabernamontana

repens fl. pleno J.Bauhin
arvensis echinatus Paulus Ammannus
prat. rad. verticilli modo rotunda C.Bauhin
tuberosus major J.Bauhin
Crus Galli Otto Brunfelsius
Coronopus parvus Batrachion Apuleius Dodonaeus (Dodoens)
Ranunculus prat. parvus fol. trifido C.Bauhin
arvensis annuus fl. minimo luteo Morison
fasciatus Henricus Volgnadius
Ol. Borrich Caspar Bartholino

These were, of course, common or vernacular names with wide currency and strong candidates for inclusion in lists which were intended to clarify the complicated state of plant naming. Local, vulgar names escaped such listing until much later times, when they were being less used and lexicographers began to collect them, saving most from vanishing for ever.

Great advances were made during the seventeenth century. Robert Morison (1620–1683) published a convenient or artificial system of grouping 'kinds' into groups of increasing size, as a hierarchy. One of his groups we now call the family *Umbelliferae* or, to give it its modern name, *Apiaceae*, and this was the first natural group to be recognized. By natural group we imply that the members of the group share a sufficient number of common features to suggest that they have all evolved from a common ancestral stock. Joseph Pitton de Tournefort (1656–1708) had made a very methodical survey of plants and had assorted 10,000 'kinds' into 698 groups (or genera). The 'kinds' must now be regarded as the basic units of classification called species. Although critical observation of structural and anatomical features led to classification advancing beyond the vague herbal and signature systems, no such advance was made in plant naming until a Swede, of little academic ability when young, we are told, established landmarks in both classification and nomenclature of plants. He was Carl Linnaeus (1707–1778), who classified

7,300 species into 1,098 genera and gave to each species a binomial name (a name consisting of a generic name-word plus a descriptive epithet, both of Latin form).

It was inevitable that, as man grouped the ever-increasing number of known plants (and he was then principally aware of those from Europe, the Mediterranean and a few from other areas) the constancy of associated morphological features in some groups should suggest that the whole was derived, by evolution, from a common ancestor. Morison's family *Umbelliferae* was a case in point. Also, because the basic unit of any system of classification is the species, and some species were found to be far less constant than others, it was just as inevitable that the nature of the species itself would become a matter of controversy, not least in terms of religious dogma. A point often passed over with insufficient comment is that Linnaeus' endeavours towards a natural system of classification were accompanied by his changing attitude towards Divine Creation. From the 365 aphorisms by which he expressed his views in *Fundamenta Botanica* (1736), and expanded in *Critica Botanica* (1737), his early view was that all species were produced by the hand of the Almighty Creator and that 'variations in the outside shell' were the work of 'Nature in a sporty mood'. In such genera as *Thalictrum* and *Clematis*, he later concluded that some species were not original creations and, in *Rosa*, he was drawn to conclude that either some species had blended or that one species had given rise to several others. Later, he invoked hybridization as the process by which species could be created and attributed to the Almighty the creation of the primeval genera, each with a single species. From his observation of land accretion during trips to Öland and Gotland, in 1741, he accepted a continuous creation of the earth and that Nature was in continuous change (*Oratio de Telluris habitabilis incremento*, 1744). He later accepted that fossil bed remains could only be explained by a process of continuous creation. In *Genera Plantarum*, 6th edn. (1764), he attributed to God the creation of the natural orders (our families). Nature produced from these the genera and species, and permanent varieties were produced by hybridization between them. The abnormal varieties of the species so formed were the product of chance.

Linnaeus was well aware of the results which plant hybridizers were obtaining in Holland and it is not surprising that his own knowledge of naturally occurring variants led him towards a covertly expressed belief in evolution. However, that expression, and his listing of varieties under their typical species in *Species Plantarum*, where he indicated each with a Greek letter, was still contrary to the dogma of Divine Creation and it would be another century before an authoritative declaration of evolutionary theory was to be made, by Charles Darwin (1809–1882).

Darwin's essay on 'The Origin of Species by Means of Natural Selection' (1859) was published somewhat reluctantly and in the face of fierce opposition. It was concerned with the major evolutionary changes by which species evolve and was based upon Darwin's own observations on fossils and living creatures. The concept of natural selection, or the survival of any life-form being dependent upon its ability to compete successfully for a place in nature, became, and still is, accepted as the major force directing an inevitable process of organic change. Our conception of the mechanisms and the causative factors for the large evolutionary steps, such as the demise of the dinosaurs and of many plant groups now known only as fossils, and the emergence and diversification of the flowering plants during the last 100 million years is, at best, hazy.

The great age of plant-hunting, from the second half of the eighteenth century through most of the nineteenth century, produced a flood of species not previously known. Strange and exotic plants were once prized above gold and caused theft, bribery and murder. Trading in 'paper tulips' by the van Bourse family gave rise to the continental stock exchange – the Bourse. With the invention of the Wardian Case by Dr Nathaniel Bagshaw Ward, in 1827, it became possible to transport plants from the farthest corners of the world by sea and without enormous losses. The case was a small glasshouse, which reduced water losses and made it unnecessary to use large quantities of fresh water on the plants during long sea voyages, as well as giving protection from salt spray. In the confusion which resulted from the naming of this flood of plants, and the use of many languages to describe them, it became apparent that there was a need for international agreement on both these matters. Today, we have rules formulated to govern the names of about 300,000 species of plants, which are now generally accepted, and have disposed of a great number of names that have been found invalid.

Our present state of knowledge about the mechanisms of inheritance and change in plants and animals is almost entirely limited to an understanding of the causes of variation within a species. That understanding is based upon the observed behaviour of inherited characters as first recorded in *Pisum* by Gregor Johann Mendel, in 1866. With the technical development of the microscope, Malpighi (1671), Grew (1672) and others explored the cellular structure of plants and elucidated the mechanism of fertilization. However, the nature of inheritance and variability remained clouded by myth and monsters until Mendel's work was rediscovered at the beginning of the twentieth century. By 1900, deVries, Correns, Tschermak and Bateson had confirmed that inheritance had a definite, particulate character which is regulated by 'genes'. Sutton (1902) was the first person to clarify the manner in which the characters are transmitted from parents to offspring when he described the behaviour of 'chromosomes' during division of the cell nucleus. Chromosomes are thread-like bodies which can be stained in dividing cells so that the sequence of events of their own division can be followed. Along their length, it can be shown, the sites of genetic control, or genes, are situated in an ordered linear sequence. Differences between individuals can now be explained in terms of the different forms, or allelomorphs, in which single genes can exist as a consequence of their mutation. At the level of the gene, we must now consider the mutants and alleles as variants in molecular structure represented by the sequences of bases in the desoxyribonucleic acid. Classification can not yet accommodate the new, genetically modified forms that may only be distinguished in terms of some property resultant upon the insertion of a fragment of DNA.

The concept of a taxonomic species, or grouping of individuals each of which has a close resemblance to the others in every aspect of its morphology, and to which a name can be applied, is not always the most accurate interpretation of the true circumstances in nature. It defines and delimits an entity but we are constantly discovering that the species is far from being an immutable entity. The botanist discovers that a species has components which have well-defined, individual ecotypic properties (an ability to live on a distinctive soil type, or an adaptation to flower and fruit in harmony with some agricultural practice) or have reproductive barriers caused by differences in chromosome number, etc. The plant breeder produces a steady stream of new varieties of cultivated species by hybridization and selection from the progeny. Genetically modified plants with very specific 'economic' properties are produced by techniques which evade nature's safeguards of incompatibility and hybrid sterility and may or may not have to be repeatedly re-synthesized.

If we consider some of the implications of, and attitudes towards, delimiting plant species and their components, and naming them, it will become easier to understand the need for internationally accepted rules intended to prevent the unnecessary and unacceptable proliferation of names.

Towards a solution to the problem

It is basic to the collector's art to arrange items into groups. Postage stamps can be arranged by country of origin and then on face value, year of issue, design, colour variation, or defects. The arranging process always resolves into a hierarchic set of groups. In the plant kingdom we have a descending hierarchy of groups through Divisions, divided into Classes, divided into Orders, divided into Families, divided into Genera, divided into Species. Subsidiary groupings are possible at each level of this hierarchy and are employed to rationalize the uniformity of relationships within the particular group. Thus, a genus may be divided into a mini-hierarchy of subgenera, divided into sections, divided into series in order to assort the components into groupings of close relatives. All such components would, nevertheless, be members of the one genus.

Early systems of classification were much less sophisticated and were based upon few aspects of plant structure such as those which suggested signatures, and mainly upon ancient herbal and medicinal concepts. Later systems would reflect advances in man's comprehension of plant structure and function, and employ the morphology and anatomy of reproductive structures as defining features. Groupings such as Natural Orders and Genera had no precise limits or absolute parity, one with another; and genera are still very diverse in size, distribution and the extent to which they have been subdivided.

Otto Brunfels (1489–1534) was probably the first person to introduce accurate, objective recording and illustration of plant structure in his *Herbarium* of 1530, and Valerius Cordus (1515–1544) could have revolutionized botany but for his premature death. His four books of German plants contained detailed accounts of the structure of 446 plants, based upon his own systematic studies on them. Many of the plants were new to science. A fifth book on Italian plants was in compilation when he died. Conrad Gesner (1516–1565) published Cordus' work on German plants in 1561 and the fifth book in 1563.

A primitive suggestion of an evolutionary sequence was contained in Matthias de l'Obel's *Plantarum seu Stirpium Historia* (1576) in which narrow-leaved plants, followed by broader-leaved, bulbous and rhizomatous plants, followed by herbaceous dicotyledons, followed by shrubs and trees, was regarded as a series of increasing 'perfection'. Andrea Caesalpino (1519–1603) retained the distinction between woody and herbaceous plants but employed more detail of flower, fruit and seed structure in compiling his classes of plants (*De Plantis*, 1583). His influence extended to the classifications of Caspar Bauhin (1550–1624), who departed from the use of medicinal information and compiled detailed descriptions of the plants to which he gave many two-word names, or binomials. P.R. de Belleval (1558–1632) adopted a binomial system which named each plant with a Latin noun followed by a Greek adjectival epithet. Joachim Jung (1587–1657) feared being accused of heresy, which prevented him from publishing his work. The manuscripts which survived him contain many of the terms which we still use in describing leaf and flower structure and arrangement, and also contain plant names consisting of a noun qualified by an adjective. Robert Morison (1620–1683) used binomials, and John Ray (1627–1705), who introduced the distinction between monocotyledons and dicotyledons, but retained the distinction between flowering herbaceous plants and woody plants, also used binomial names.

Joseph Pitton de Tournefort (1656–1708) placed great emphasis on the floral corolla and upon defining the genus, rather than the species. His 698 generic descriptions are detailed but his species descriptions are dependent upon binomials and illustrations. Herman Boerhaave (1668–1739) combined the systems of Ray and Tournefort, and others, to incorporate morphological, ecological, leaf, floral and fruiting characters, but none of these early advances received popular support. As Michel Adanson (1727–1806) was to realize, some sixty systems of classification had been proposed by the middle of the eighteenth century and none had been free from narrow conceptual restraints. His plea that attention should be focused on 'natural' classification through processes of inductive reasoning, because of the wide range of characteristics then being employed, did not enjoy wide publication and his work was not well regarded when it did become more widely known. His main claim to fame, or notoriety, stems from his use of names which have no meanings.

Before considering the major contributions made by Carl Linnaeus, it should be noted that the names of many higher groups of plants, of families and of genera were well established at the beginning of the eighteenth century and several people had used simplified, binomial names for species. Indeed, August Quirinus Rivinus (1652–1723) had proposed that no plant should have a name of more than two words.

Carl Linnaeus (1707–1778) was the son of a clergyman, Nils, who had adopted the latinized family name when he became a student of theology. Carl also went to theological college for a year but then left and became an assistant gardener in Prof. Olof Rudbeck's botanic garden at Uppsala. His ability as a collector and arranger soon became evident and, after undertaking tours through Lapland, he began to publish works which are now the starting points for naming plants and animals. In literature he is referred to as Carl or Karl or Carolus Linnaeus, Carl Linné (an abbreviation) and, later in life, as Carl von Linné. His life became one of devotion to the classification and naming of all living things and of teaching others about them. His numerous students played a very important part in the discovery of new plants from many parts of the world. Linnaeus' main contribution to botany was his method of naming plants, in which he combined Bauhin's and Belleval's use of binomials with Tournefort's and Boerhaave's concepts of the genus. His success, where others before him had failed, was due to the early publication of his most popular work, an artificial system of classifying plants. In this he employed the number, structure and disposition of the stamens of the flower to define 23 classes, each subdivided into orders on the basis of the number of parts constituting the pistil, with a 24th class containing those plants which had their reproductive organs hidden to the eye: the orders of which were the ferns, mosses, algae (in which he placed liverworts, lichens and sponges), fungi and palms. This 'sexual system' provided an easy way of grouping plants and of allocating newly discovered plants to a group. Originally designed to accommodate the plants of his home parish, it was elaborated to include first the Arctic flora and later the more diverse and exotic plants being discovered in the tropics. It continued in popular use into the nineteenth century despite its limitation of grouping together strange bedfellows: red valerian, tamarind, crocus, iris, galingale sedge and mat grass are all grouped under *Triandria* (three stamens) *Monogynia* (pistil with a single style).

In 1735, Linnaeus published *Systema Naturae*, in which he grouped species into genera, genera into orders and orders into classes on the basis of structural similarities. This was an attempt to interpret evolutionary relationships or assemblages of individuals at different levels. It owed much to a collaborator and fellow student of Linnaeus, Peter Artendi (d. 1735) who, before an untimely death, was working on the classification of fishes, reptiles and amphibians, and the *Umbelliferae*. In *Species Plantarum*, published in 1753, Linnaeus gave each species a binomial name. The first word of each binomial was the name of the genus to which the species belonged and the second word was a descriptive, or specific epithet. Both words were in Latin or Latin form. Thus, the creeping buttercup he named as *Ranunculus repens*.

It now required that the systematic classification and the binomial nomenclature, which Linnaeus had adopted, should become generally accepted and, largely because of the popularity of his sexual system, this was to be the case. Botany could now contend with the rapidly increasing number of species of plants being collected for scientific enquiry, rather than for medicine or exotic gardening, as in the seventeenth century. For the proper working of such standardized nomenclature, however, it was necessary that the language of plant names should also be standardized. Linnaeus' views on the manner of forming plant names, and the use of Latin for these and for the descriptions of plants and their parts, have given rise directly to modern practice and a Latin vocabulary of great versatility, but which would have been largely incomprehensible in ancient Rome. He applied the same methodical principles to the naming of animals, minerals and diseases and, in doing so, established Latin, which was the *lingua franca* of his day, as the internationally used language of science and medicine.

The rules by which we now name plants depend largely on Linnaeus' writings but, for the names of plant families, we are much dependent on A.L. de Jussieu's classification in his *Genera Plantarum* of 1789. For the name of a species, the correct name is that which was first published since 1753. This establishes Linnaeus' *Species Plantarum* (associated with his *Genera Plantarum*, 5th edn. 1754 and 6th edn. 1764) as the starting point for the names of species (and their descriptions). Linnaeus' sexual system of classification was very artificial and, although Linnaeus must have been delighted at its popularity, he regarded it as no more than a convenient pigeonholing system. He published some of his views on grouping plant genera into natural orders (our families) in *Philosophia Botanica* (1751). Most of his orders were not natural groupings but considerably mixed assemblages. By contrast, Bernard de Jussieu (1699–1777), followed by his nephew Antoine Laurent de Jussieu (1748–1836), searched for improved ways of arranging and grouping plants as natural groups. In A.L. de Jussieu's *de Genera Plantarum* (1789) the characteristics are given for 100 plant families; and most of these we still recognize.

Augustin Pyrame de Candolle (1778–1841) also sought a natural system, as did his son Alphonse, and he took the evolutionist view that there is an underlying state of symmetry in the floral structure which we can observe today and that, by considering relationships in terms of that symmetry, natural alliances may be recognized. This approach resulted in a great deal of monographic work from which de Candolle formed views on the concept of a core of similarity, or type, for any natural group and the requirement for control in the naming of plants.

Today, technological and scientific advances have made it possible for us to use subcellular, chemical and the minutest of morphological features and to incorporate as many items of information as are available about a plant in computer-aided assessments of that plant's relationships to others. Biological information has often been found to conflict with the concept of the taxonomic species and there are many plant groups in which the 'species' can best be regarded as a collection of highly variable populations. The gleaning of new evidence necessitates a continuing process of reappraisal of families, genera and species. Such reappraisal may result in subdivision or even splitting of a group into several new ones or, the converse process, in lumping together two or more former groups into one new one. Since the bulk of research is carried out on the individual species, most of the revisions are carried out at or below the rank of species. On occasion, therefore, a revision at the family level will require the transfer of whole genera from one family to another, but it is now more common for a revision at the level of the genus to require the transfer of some, if not all the species from one genus to another. Such revisions are not mischievous but are the necessary process by which newly acquired knowledge is incorporated into a generally accepted framework. It is because we continue to improve the extent of our knowledge of plants that revision of the systems for their classification continues and, consequently, that name changes are inevitable.

The equivalence, certainly in evolutionary terms, of groups of higher rank than of family is a matter of philosophical debate and, even at the family level, we find divergence of views as to whether those with few components are equivalent to those with many components. Over the past twenty years a 'Family Planning' committee of taxonomists has met in London to determine an acceptable system of plant families in view of the variation presented by systematists since Bentham and Hooker. In the petaloid monocotyledons they were unanimous in agreeing to split the lilies (essentially the familiar families *Liliaceae* and *Amaryllidaceae*) to make the family concept more comparable with that adopted in other groups. The following liliaceous family names are now in common use: *Melanthiaceae*, *Colchicaceae*, *Asphodelaceae*, *Hyacinthaceae*, *Hemerocallidaceae*, *Agavaceae*, *Aphyllandraceae*, *Lomandraceae*, *Anthericaceae*, *Xanthorrhoeaceae*, *Alliaceae*, *Liliaceae*, *Dracaenaceae*, *Asparagaceae*, *Ruscaceae*, *Convallariaceae*, *Trilliaceae*, *Alteriaceae*, *Herreriaceae*, *Philesiaceae*, *Smilacaceae*, *Haemadoraceae*, *Hypoxidaceae*, *Alstoemeriaceae*, *Doryanthaceae*, *Campynemaceae* and *Amaryllidaceae*.

Because the taxonomic species is the basic unit of any system of classification, we have to assume parity between species; that is to say, we assume that a widespread species is in every way comparable with a rare species which may be restricted in its distribution to a very small area. It is a feature of plants that their diversity – of habit, longevity, mode of reproduction and tolerance of environmental conditions – presents a wide range of biologically different circumstances. For the taxonomic problem of delimiting, defining and naming a species we have to identify a grouping of individuals whose characteristics are sufficiently stable to be defined, in order that a name can be applied to the group and a 'type', or exemplar, can be specified for that name. It is because of this concept of the 'type' that changes have to be made in names of species in the light of new discoveries and that entities below the rank of species have to be recognized. Thus, we speak of a botanical 'sub-species' when part of the species grouping can be distinguished as having a number of features which remain constant and as having a distinctive geographical or ecological distribution. When the degree of departure from the typical material is of a lesser order we may employ the inferior category of 'variety'. The term 'form' is employed to describe a variant which is distinct in a minor way only, such as a single feature difference which might appear sporadically due to genetic mutation or sporting.

The patterns and causes of variation differ from one species to another and this has long been recognized as a problem in fully reconciling the idea of a taxonomic species with that of a biological system of populations in perpetual evolutionary flux. Below the level of species, agreement about absolute ranking is far from complete and even the rigidity of the infraspecific hierarchy (*subspecies*, *varietas*, *subvarietas*, *forma*, *subforma*) is now open to question.

It is always a cause of annoyance when a new name has to be given to a plant which is widely known under its superseded old name. Gardeners always complain about such name changes but there is no novelty in that. On the occasion of Linnaeus being proposed for Fellowship of the Royal Society, Peter Collinson wrote to him in praise of his *Species Plantarum* but, at the same time, complained that Linnaeus had introduced new names for so many well-known plants.

The gardener has some cause to be aggrieved by changes in botanical names. Few gardeners show much alacrity in adopting new names and perusal of gardening books and catalogues shows that horticulture seldom uses botanical names with all the exactitude which they can provide. Horticulture, however, not only agreed to observe the international rules of botanical nomenclature but also formulated its own additional rules for the naming of plants grown under cultivation. It might appear as though the botanist realizes that he is bound by the rules, whereas the horticulturalist does not, but to understand this we must recognize the different facets of horticulture. The rules are of greatest interest and importance to specialist plant breeders and gardeners with a particular interest in a certain plant group. For the domestic gardener it is the growing of beautiful plants which is the motive force behind his activity. Between the two extremes lies every shade of interest and the main emphasis on names is an emphasis on garden names. Roses, cabbages, carnations and leeks are perfectly adequate names for the majority of gardeners but if greater precision is needed, a gardener wishes to know the name of the variety. Consequently, most gardeners are satisfied with a naming system which has no recourse to the botanical rules whatsoever. Not surprisingly, therefore, seed and plant catalogues also avoid botanical names. The specialist plant breeder, however, shows certain similarities to the apothecaries of an earlier age. Like them, he guards his art and his plants jealously because they represent the source of his future income and, also like them, he has the desire to understand every aspect of his plants. The apothecaries gave us the first centres of botanical enquiry and the plant breeders of today give us the new varieties which are needed to satisfy our gardening and food-production requirements. The commercial face of plant breeding, however, attaches a powerful monetary significance to the names given to new varieties.

Gardeners occasionally have to resort to botanical names when they discover some cultural problem with a plant which shares the same common name with several different plants. The Guernsey lily, around which has always hung a cloud of mystery, has been offered to the public in the form of *Amaryllis belladonna* L. The true Guernsey lily has the name *Nerine sarniensis* Herb. (but was named *Amaryllis sarniensis* by Linnaeus). The epithet *sarniensis* means 'of Sarnia' or 'of Guernsey', Sarnia being the old name for Guernsey, and is an example of a misapplied geographical epithet, since the plant's native area is South Africa. Some would regard the epithet as indicating the fact that Guernsey was the first place in which the plant was cultivated. This is historically incorrect, however, and does nothing to help the gardener who finds that the Guernsey lily that he has bought does not behave, in culture, as *Nerine sarniensis* is known to behave. This example is one involving a particularly contentious area as to the taxonomic problems of generic boundaries and typification but there are many others in which common and Latin garden names are used for whole assortments of garden plants, ranging from species (*Nepeta mussinii* and *N. cataria* are both catmint) to members of different genera ('japonicas' including *Chaenomeles speciosa* and *Kerria japonica*) to members of different families (*Camellia japonica* is likewise a 'japonica'), and the diversity of 'bluebells' was mentioned earlier.

New varieties, be they timber trees, crop plants or garden flowers, require names and those names need to be definitive. As with the earlier confusion of botanical names (different names for the same species or the same name for different species), so there can be the same confusion of horticultural names. As will be seen, rules for cultivated plants require that new names have to be established by publication. This gives to the breeder the commercial advantage of being able to supply to the public his new variety under what, initially, amounts to his mark of copyright. In some parts of the world legislation permits exemption from the rules and recommendations otherwise used for the names of cultivated plants.

The rules of botanical nomenclature

The rules which now govern the naming and the names of plants really had their beginnings in the views of A.P. de Candolle as he expressed them in his *Théorie Elémentaire de la Botanique* (1813). There, he advised that plants should have names in Latin (or Latin form but not compounded from different languages), formed according to the rules of Latin grammar and subject to the right of priority for the name given by the discoverer or the first describer. This advice was found inadequate and, in 1862, the International Botanical Congress in London adopted control over agreements on nomenclature. Alphonse de Candolle (1806–1893), who was A.P. de Candolle's son, drew up four simple 'Lois', or laws, which were aimed at resolving what threatened to become a chaotic state of plant nomenclature. The Paris International Botanical Congress of 1867 adopted the Lois, which were:

1 One plant species shall have no more than one name.
2 No two plant species shall share the same name.
3 If a plant has two names, the name which is valid shall be that which was the earliest one to be published after 1753.
4 The author's name shall be cited, after the name of the plant, in order to establish the sense in which the name is used and its priority over other names.

It can be seen from the above Lois that, until the nineteenth century, botanists frequently gave names to plants with little regard either to the previous use of the same name or to names that had already been applied to the same plant. It is because of this aspect that one often encounters the words *sensu* and *non* inserted before the name of an author, although both terms are more commonly used in the sense of taxonomic revision, and indicate that the name is being used 'in the sense of' or 'not in the sense of' that author, respectively.

The use of Latin, as the language in which descriptions and diagnoses were written, was not universal in the nineteenth century and many regional languages were used in different parts of the world. A description is an account of the plant's habit, morphology and periodicity whereas a diagnosis is an author's definitive statement of the plant's diagnostic features, and circumscribes the limits outside which plants do not pertain to that named species. A diagnosis often states particular ways in which the species differs from another species of the same genus. Before the adoption of Latin as the accepted language of botanical nomenclature, searching for names already in existence for a particular plant, and confirming their applicability, involved searching through multilingual literature. The requirement to use Latin was written into the rules by the International Botanical Congress in Vienna, in 1905. However, the American Society of Plant Taxonomists produced its own Code in 1947, which became known as the Brittonia edition of the Rules or the Rochester Code, and disregarded this requirement. Not until 1959 was international agreement achieved and then the requirement to use Latin was made retroactive to January 1st, 1935, the year of the Amsterdam meeting of the Congress.

The rules are considered at each International Botanical Congress, formerly held at five-, and more recently at six-, yearly intervals during peacetime. The International Code of Botanical Nomenclature (first published as such in 1952) was formulated at the Stockholm Congress of 1950. In 1930, the matter of determining the priority of specific epithets was the main point at issue. The practice of British botanists had been to regard that epithet which was first published after the plant had been allocated to its correct genus as the correct name. This has been called the Kew Rule, but it was defeated in favour of the rule that now gives priority to the epithet that was the first to be published from the starting date of

May 1st, 1753. Epithets which predate the starting point, but which were adopted by Linnaeus, are attributed to Linnaeus (e.g. Bauhin's *Alsine media*, *Ammi majus*, *Anagyris foetida* and *Galium rubrum* and Dodoens' *Angelica sylvestris* are examples of binomials nevertheless credited to Linnaeus).

The 1959 International Botanical Congress in Montreal introduced the requirement under the Code, that for valid publication of a name of a family or any taxon of lower rank, the author of that name should cite a 'type' for the name and that this requirement should be retrospective to January 1st, 1958. The idea of a type goes back to A.P. de Candolle and it implies a representative collection of characteristics to which a name applies. The type in Botany is a nomenclatural type: it is the type for the name and the name is permanently attached to it or associated with it. For the name of a family, the representative characteristics which that name implies are those embodied in one of its genera, which is called the type genus. In a similar way, the type for the name of a genus is the type species of that genus. For the name of a species or taxon of lower rank, the type is a specimen lodged in an herbarium or, in certain cases, published illustrations. The type need not, nor could it, be representative of the full range of entities to which the name is applied. Just as a genus, although having the features of its parent family, cannot be fully representative of all the genera belonging to that family, no single specimen can be representative of the full range of variety found within a species.

For the name to become the correct name of a plant or plant group, it must satisfy two sets of conditions. First, it must be constructed in accordance with the rules of name formation, which ensures its legitimacy. Second, it must be published in such a way as to make it valid. Publication has to be in printed matter which is distributed to the general public or, at least, to botanical institutions with libraries accessible to botanists generally. Since January 1st, 1953, this has excluded publication in newspapers and tradesmen's catalogues. Valid publication also requires the name to be accompanied by a description or diagnosis, an indication of its rank and the nomenclatural type, as required by the rules. This publication requirement, and subsequent citation of the new name followed by the name of its author, ensures that a date can be placed upon the name's publication and that it can, therefore, be properly considered in matters of priority.

The present scope of the Code is expressed in the Principles, which have evolved from the de Candollean Lois:

1 Botanical nomenclature is independent of zoological nomenclature. The Code applies equally to names of taxonomic groups treated as plants whether or not these groups were originally so treated.
2 The application of names of taxonomic groups is determined by means of nomenclatural types.
3 The nomenclature of a taxonomic group is based upon priority of publication.
4 Each taxonomic group with a particular circumscription, position and rank can bear only one correct name, the earliest which is in accordance with the rules, except in specified cases.
5 Scientific names of taxonomic groups are treated as Latin, regardless of their derivation.
6 The rules of nomenclature are retroactive unless expressly limited.

The detailed rules are contained in the Articles and Recommendations of the Code and mastery of these can only be gained by practical experience (Greuter, 2000). A most lucid summary and comparison with other Codes of biological nomenclature is that of Jeffrey (1978), written for the Systematics Association.

There are still new species of plants to be discovered and an enormous amount of information yet to be sought for long-familiar species, in particular, evidence of a chemical nature, and especially that concerned with proteins, which may provide reliable indications of phylogenetic relationships. For modern systematists, the greatest and most persistent problem is our ignorance about the apparently explosive appearance of a diverse array of flowering plants, some 100 million years ago, from one or more unknown ancestors. Modern systems of classification are still frameworks within which the authors arrange assemblages in sequences or clusters to represent their own idiosyncratic interpretation of the known facts. In addition to having no firm record of the early evolutionary pathways of the flowering plants, the systematist also has the major problems of identifying clear-cut boundaries between groups and of assessing the absolute ranking of groups. It is because of these continuing problems that, although the Code extends to taxa of all ranks, most of the rules are concerned with the names and naming of groups from the rank of family downwards.

Before moving on to the question of plant names at the generic and lower ranks, this is a suitable point at which to comment on new names for families which are now starting to appear in books and catalogues, and some explanation in passing may help to dispel any confusion. The splitting of the *Liliaceae* and *Amaryllidaceae* into 27 new families was mentioned on page 21 but the move towards standardization has required other family name changes.

Family names

Each family can have only one correct name and that, of course, is the earliest legitimate one, except in cases of limitation of priority by conservation. In other words, there is provision in the Code for disregarding the requirement of priority when a special case is proved for a name to be conserved. Conservation of names is intended to avoid disadvantageous name changes, even though the name in question does not meet all the requirements of the Code. Names which have long-standing use and wide acceptability and are used in standard works of literature can be proposed for conservation and, when accepted, need not be discarded in favour of new and more correct names.

The names of families are plural adjectives used as nouns and are formed by adding the suffix -aceae to the stem, which is the name of an included genus. Thus, the buttercup genus *Ranunculus* gives us the name *Ranunculaceae* for the buttercup family and the water-lily genus *Nymphaea* gives us the name *Nymphaeaceae* for the water-lilies. A few family names are conserved, for the reasons given above, which do have generic names as their stem, although one, the *Ebenaceae*, has the name *Ebenus* Kuntze (1891) *non* Linnaeus (1753) as its stem. Kuntze's genus is now called *Maba* but its parent family retains the name *Ebenaceae* even though *Ebenus* L. is the name used for a genus of the pea family. There are eight families for which specific exceptions are provided and which can be referred to either by their long-standing, conserved names or, as is increasingly the case in recent floras and other published works on plants, by their names which are in agreement with the Code. These families and their equivalent names are:

Compositae	or	*Asteraceae* (on the genus *Aster*)
Cruciferae	or	*Brassicaceae* (on the genus *Brassica*)
Gramineae	or	*Poaceae* (on the genus *Poa*)
Guttiferae	or	*Clusiaceae* (on the genus *Clusia*)
Labiatae	or	*Lamiaceae* (on the genus *Lamium*)
Leguminosae	or	*Fabaceae* (on the genus *Faba*)
Palmae	or	*Arecaceae* (on the genus *Areca*)
Umbelliferae	or	*Apiaceae* (on the genus *Apium*)

Some botanists regard the *Leguminosae* as including three subfamilies but others accept those three components as each having family status. In the latter case, the three families are the *Caesalpiniaceae*, the *Mimosaceae* and the *Papilionaceae*. The last of these family names refers to the resemblance which may be seen in the pea- or bean-flower structure, with its large and colourful sail petal, to a resting butterfly (*Papilionoidea*) and is not based upon the name of a plant genus. If a botanist wishes to retain the three-family concept, the name *Papilionaceae* is conserved against *Leguminosae* and the modern equivalent is *Fabaceae*. Consequently, the *Fabaceae* are either the entire aggregation of leguminous plant genera or that part of the aggregate which does not belong in either the *Caesalpiniaceae* or the *Mimosaceae*.

Some eastern European publications use *Daucaceae* for the *Apiaceae*, split the *Asteraceae* into *Carduaceae* and *Chicoriaceae* and adopt various views as to the generic basis of family names (e.g. *Oenotheraceae* for *Onagraceae* by insisting that Linnaeus' genus *Oenothera* has prior claim over Miller's genus *Onagra*).

Generic names

The name of a genus is a noun, or word treated as such, and begins with a capital letter. It is singular, may be taken from any source whatever, and may even be composed in an arbitrary manner. The etymology of generic names is, therefore, not always complete and, even though the derivation of some may be discovered, they lack meaning. By way of examples:

Portulaca, from the Latin *porto* (I carry) and *lac* (milk) translates as 'Milk-carrier'.

Pittosporum, from the Greek, πιττοω (I tar) and σπορος (a seed) translates as 'Tar-seed'.

Hebe was the goddess of youth and, amongst other things, the daughter of Jupiter. It cannot be translated further.

Petunia is taken from the Brazilian name for tobacco.

Tecoma is taken from a Mexican name.

Linnaea is one of the names which commemorate Linnaeus.

Sibara is an anagram of *Arabis*.

Aa is the name given by Reichenbach to an orchid genus which he segregated from *Altensteinia*. It has no meaning and, as others have observed, must always appear first in an alphabetic listing.

The generic names of some Old World plants were taken from Greek mythology by the ancients, or are identical to the names of characters in Greek mythology. The reason for this is not always clear (e.g. *Althaea*, *Cecropia*, *Circaea*, *Melia*, *Phoenix*, *Tagetes*, *Thalia*, *Endymion*, *Hebe*, *Paeonia* and *Paris*). However, some do have reasonable floristic associations, e.g. *Atropa* (the third Fate, who held the scissors to cut the thread of life), *Chloris* (the Goddess of flowers), *Iris* (messenger to Gods of the rainbow), *Melissa* (apiarist who used the plant to feed the bees). The metamorphoses, that are so common in the mythology, provided direct associations for several names, e.g. *Acanthus* (became an *Acanthus*), *Adonis* (became an *Anemone*), *Ajacis* (became a *Narcissus*), *Daphne* (became a laurel), *Hyacinthus* (became, probably, a *Delphinium*) and *Narcissus* (became a daffodil).

If all specific names were constructed in the arbitrary manner used by M. Adanson (1727–1806), there would have been no enquiries of the author and this book would not have been written. In fact, the etymology of plant names is a rich store of historical interest and conceals many facets of humanity ranging from the sarcasm of some authors to the humour of others. This is made possible by the wide scope available to authors for formulating names and because, whatever language is the source, names are treated as being in Latin. Imaginative association has produced some names which are very descriptive provided that the reader can spot the association. In the algae, the Chrysophyte which twirls like a ballerina has been named *Pavlova gyrans* and, in the fungi, a saprophyte on leaves of *Eucalyptus* which has a wide-mouthed spore-producing structure has been named *Satchmopsis brasiliensis* (Satchmo, satchelmouth). The large vocabulary of botanical Latin comes mostly from the Greek and Latin of ancient times but, since the ancients had few words which related specifically to plants and their parts, a Latin dictionary is of somewhat limited use in trying to decipher plant diagnoses. By way of examples, Table 1 gives the parts of the flower (Latin *flos*, Greek ανθος) (illustrated in Fig. 1) and the classical words from which they are derived, together with their original sense.

The grammar of botanical Latin is very formal and much more simple than that of the classical language itself. A full and most authoritative work on the subject is contained in Stearn's book,

Table 1

Flower part	Greek	Latin	Former meaning
calyx	κάλυξ	—	various kinds of covering
	κύλίξ	—	cup or goblet
sepal	σκέπη	—	covering
corolla	—	*corolla*	garland or coronet
petal	πέταλον	—	leaf
	—	*petalum*	metal plate
stamen	—	*stamen*	thread, warp, string
filament	—	*filamentum*	thread
anther	—	*anthera*	potion of herbs
androecium	ἀυδρ-, οἰκός	—	man-, house
stigma	στίγμα	—	tattoo or spot
style	στῦλος	—	pillar or post
	—	*stilus*	pointed writing tool
carpel	καρπός	—	fruit
gynoecium	γυνή-, οἰκός	—	woman-, house
pistil	—	*pistillum*	pestle

Botanical Latin (1983). Nevertheless, it is necessary to know that in Latin, nouns (such as family and generic names) have gender, number and case and that the words which give some attribute to a noun (as in adjectival specific epithets) must agree with the noun in each of these. Having gender means that all things (the names of which are called nouns) are either masculine or feminine or neuter. In English, we treat almost everything as neuter, referring to nouns as 'it', except animals and most ships and aeroplanes (which are commonly held to be feminine). Gender is explained further below. Number means that things may be single (singular) or multiple (plural). In English we either have different words for the singular and plural (man and men, mouse and mice) or we convert the singular into the plural most commonly by adding an 's' (ship and ships, rat and rats) or more rarely by adding 'es' (box and boxes, fox and foxes) or, rarer still, by adding 'en' (ox and oxen). In Latin, the difference is expressed by changes in the endings of the words. Case is less easy to understand but means the significance of the noun to the meaning of the sentence in which it is contained. It is also expressed in the endings of the words. In the sentence, 'The flower has charm', the flower is singular, is the subject of the sentence and has what is called the nominative case. In the sentence 'I threw away the flower', I am now the subject and the flower has become the direct object in the accusative case. In the sentence, 'I did not like the colour of the flower', I am again the subject, the colour is now the object and the flower has become a possessive noun and has the genitive case. In the sentence, 'The flower fell to the ground' the flower is once again the subject (nominative) and the ground has the dative case. If we add 'with a whisper', then whisper takes the ablative case. In other words, case confers on nouns an expression of their meaning in any sentence. This is shown by the ending of the Latin word, which changes with case and number and, in so doing, changes the naked word into part of a sentence (Table 2).

Nouns fall into five groups, or declensions, as determined by their endings (Table 3).

Generic names are treated as singular subjects, taking the nominative case. *Solanum* means 'Comforter' and derives from the use

Table 2

Case	Singular		Plural	
nominative	*flos*	the flower (subject)	*flores*	the flowers
accusative	*florem*	the flower (object)	*flores*	the flowers
genitive	*floris*	of the flower	*florum*	of the flowers
dative	*flori*	to or for the flower	*floribus*	to or for the flowers
ablative	*flore*	by, with or from the flower	*floribus*	by, with or from the flowers

Table 3

Declension		I	II		III				IV		V
Gender		f	m	n	m,f	n	m,f	n	m	n	f
Singular	nom	*-a*	*-us(-er)*	*-um*	*	*	*-is(es)*	*-e(l)(r)*	*-us*	*-u*	*-es*
	acc	*-am*	*-um*	*-um*	*-em*	*	*-em(im)*	*-e(l)(r)*	*-um*	*-u*	*-em*
	gen	*-ae*	*-i*	*-i*	*-is*	*-is*	*-is*	*-is*	*-us*	*-us*	*-ei*
	dat	*-ae*	*-o*	*-o*	*-i*	*-i*	*-i*	*-i*	*-ui(u)*	*-ui(u)*	*-ei*
	abl	*-a*	*-o*	*-o*	*-e*	*-e*	*-i(e)*	*-i(e)*	*-u*	*-u*	*-e*
Plural	nom	*-ae*	*-i*	*-a*	*-es*	*-a*	*-es*	*-ia*	*-us*	*-ua*	*-es*
	acc	*-as*	*-os*	*-a*	*-es*	*-a*	*-es(is)*	*-ia*	*-us*	*-ua*	*-es*
	gen	*-arum*	*-orum*	*-orum*	*-um*	*-um*	*-ium*	*-ium*	*-uum*	*-uum*	*-erum*
	dat	*-is*	*-is*	*-is*	*-ibus*	*-ibus*	*-ibus*	*-ibus*	*-ibus*	*-ibus*	*-ebus*
	abl	*-is*	*-is*	*-is*	*-ibus*	*-ibus*	*-ibus*	*-ibus*	*-ibus*	*-ibus*	*-ebus*

*Denotes various irregular endings.

of nightshades as herbal sedatives. The gender of generic names is that of the original Greek or Latin noun or, if that was variable, is chosen by the author of the name. There are exceptions to this in which masculine names are treated as feminine, and fewer in which compound names, which ought to be feminine, are treated as masculine. As a general guide, names ending in *-us* are masculine unless they are trees (such as *Fagus*, *Pinus*, *Quercus*, *Sorbus* which are treated as feminine), names ending in *-a* are feminine and names ending in *-um* are neuter; names ending in *-on* are masculine unless they can also take *-um*, when they are neuter, or the ending is *-dendron* when they are also neuter (*Rhododendron* or *Rhododendrum*); names ending in *-ma* (as in terminations such as *-osma*) are neuter; names ending in *-is* are mostly feminine or masculine treated as feminine (*Orchis*) and those ending in *-e* are neuter; other feminine endings are *-ago*, *-odes*, *-oides*, *-ix* and *-es*.

A recommendation for forming generic names to commemorate men or women is that these should be treated as feminine and formed as follows:

for names ending in a vowel,	terminate with *-a*
for names ending in -a,	terminate with *-ea*
for names ending in -ea,	do not change
for names ending in a consonant,	add *-ia*
for names ending in -er,	add *-a*
for latinized names ending in -us,	change the ending to *-ia*

Generic names which are formed arbitrarily or are derived from vernacular names have their ending selected by the name's author.

Species names

The name of a species is a binary combination of the generic name followed by a specific epithet. If the epithet is of two words they must be joined by a hyphen or united into one word. The epithet can be taken from any source whatever and may be constructed

Table 4

Masculine	Feminine	Neuter	Example	Meaning
-us	*-a*	*-um*	*hirsutus*	(hairy)
-is	*-is*	*-e*	*brevis*	(short)
-os	*-os*	*-on*	*acaulos* ἄκαυλος	(stemless)
-er	*-era*	*-erum*	*asper*	(rough)
-er	*-ra*	*-rum*	*scaber*	(rough)
-ax	*-ax*	*-ax*	*fallax*	(false)
-ex	*-ex*	*-ex*	*duplex*	(double)
-ox	*-ox*	*-ox*	*ferox*	(very prickly)
-ans	*-ans*	*-ans*	*reptans*	(creeping)
-ens	*-ens*	*-ens*	*repens*	(creeping)
-or	*-or*	*-or*	*tricolor*	(three-coloured)
-oides	*-oides*	*-oides*	*bryoides* βρύον, εἶδος	(moss-like)

Table 5

Masculine	Feminine	Neuter	Example	Meaning
-us	*-a*	*-um*	*longus*	(long)
-ior	*-ior*	*-ius*		(longer)
-issimus	*-issima*	*-issimum*		(longest)
-is	*-is*	*-e*	*gracilis*	(slender)
-ior	*-ior*	*-ius*		(slenderer)
-limus	*-lima*	*-limum*		(slenderest)
-er	*-era*	*-erum*	*tener*	(thin)
-erior	*-erior*	*-erius*		(thinner)
-errimus	*-errima*	*-errimum*		(thinnest)

in an arbitrary manner. It would be reasonable to expect that the epithet should have a descriptive purpose, and there are many which do, but large numbers either refer to the native area in which the plant grows or commemorate a person (often the discoverer, the introducer into cultivation or a noble personage). The epithet may be adjectival (or descriptive), qualified in various ways with prefixes and suffixes, or a noun.

It will become clear that because descriptive, adjectival epithets must agree with the generic name, the endings must change in gender, case and number; *Dipsacus fullonum* L. has the generic name used by Dioscorides meaning 'Dropsy', alluding to the accumulation of water in the leaf-bases, and an epithet which is the masculine genitive plural of *fullo*, a fuller, and which identifies the typical form of this teasel as the one which was used to clean and comb up a 'nap' on cloth. The majority of adjectival epithet endings are as in the first two examples listed in Table 4.

Comparative epithets are informative because they provide us with an indication of how the species contrasts with the general features of other members of the genus (Table 5).

Epithets commemorating people

Specific epithets which are nouns are grammatically independent of the generic name. *Campanula trachelium* is literally 'Little bell' (feminine) 'neck' (neuter). When they are derived from the names of people, they can either be retained as nouns in the genitive case (*clusii* is the genitive singular of Clusius, the latinized version of l'Ecluse, and gives an epithet with the meaning 'of l'Ecluse') or be treated as adjectives and then agreeing in gender with the generic noun (*Sorbus leyana* Wilmott is a tree taking, like many others, the feminine gender despite the masculine ending, and so the epithet which commemorates Augustin Ley also takes the feminine ending). The epithets are formed as follows:

to names ending with a vowel (except *-a*) or *-er* is added

i when masculine singular,
ae when feminine singular,
orum when masculine plural,
arum when feminine plural

to names ending with -*a* is added
e when singular,
rum when plural
to names ending with a consonant (except -*er*) is added
ii when masculine singular,
iae when feminine singular,
iorum when masculine plural,
iarum when feminine plural
or, when used adjectivally:
to names ending with a vowel (except -*a*) is added
anus when masculine,
ana when feminine,
anum when neuter
to names ending with -*a* is added
nus when masculine,
na when feminine,
num when neuter
to names ending with a consonant is added
ianus when masculine,
iana when feminine,
ianum when neuter.

Geographical epithets

When an epithet is derived from the name of a place, usually to indicate the plant's native area but also, sometimes, to indicate the area or place from which the plant was first known or in which it was produced horticulturally, it is preferably adjectival and takes one of the following endings:

-*ensis* (m)	-*ensis* (f)	-*ense* (n)
-(*a*)*nus* (m)	-(*a*)*na* (f)	-(*a*)*num* (n)
-*inus* (m)	-*ina* (f)	-*inum* (n)
-*icus* (m)	-*ica* (f)	-*icum* (n)

Geographical epithets are sometimes inaccurate because the author of the name was in error as to the true origin of the plant, or obscure because the ancient classical names are no longer familiar to us. As with epithets which are derived from proper names to commemorate people, or from generic names or vernacular names which are treated as being Latin, it is now customary to start them with a small initial letter but it remains permissible to give them a capital initial.

Categories below the rank of species

The subdivision of a species group is based upon a concept of infraspecific variation which assumes that, in nature, evolutionary changes are progressive fragmentations of the parent species. Put in another way, a species, or any taxon of lower rank, is a closed grouping whose limits embrace all their lower-ranked variants (subordinate taxa). It will be seen later that a different concept underlies the naming of cultivated plants which does not make such an assumption but recognizes the possibility that cultivars may straddle species, or other, boundaries or overlap each other, or be totally contained, one by another.

The rules by which botanical infraspecific taxa are named specify that the name shall consist of the name of the parent species followed by a term which denotes the rank of the subdivision, and an epithet which is formed in the same ways as specific epithets, including grammatical agreement when adjectival. Such names are subject to the rules of priority and typification. The ranks concerned are *subspecies* (abbreviated to *subsp.* or *ssp.*), *varietas* (variety in English, abbreviated to *var.*), *subvarietas* (subvariety or *subvar.*), *forma* (form or *f.*). These form a hierarchy and further subdivisions are permitted but the Code does not define the characteristics of any rank within the hierarchy. Consequently, infraspecific classification is subjective.

When a subdivision of a species is named, which does not include the nomenclatural type of the species, it automatically establishes the name of the equivalent subdivision which does contain that type. Such a name is an 'autonym' and has the same epithet as the species itself but is not attributed to an author. This is the only event which permits the repetition of the specific epithet and the only permissible way of indicating that the taxon includes the type for the species name. The same constraints apply to subdivisions of lower ranks. For example, *Veronica hybrida* L. was deemed by E.F. Warburg to be a component of *Veronica spicata* L. and he named it *V. spicata* L. *subsp. hybrida* (L.) E.F. Warburg. This implies the existence of a typical subspecies, the autonym for which is *V. spicata* L. *subsp. spicata*.

It will be seen from the citation of Warburg's new combination that the disappearance of a former Linnaean species can be explained. Retention of the epithet '*hybrida*', and the indication of Linnaeus being its author (in parentheses) shows the benefit of this system in constructing names with historic meanings.

Hybrids

Hybrids are particularly important as cultivated plants but are also a feature of many plant groups in the wild, especially woody perennials such as willows. The rules for the names and naming of hybrids are contained in the Botanical Code but are equally applicable to cultivated plant hybrids.

For the name of a hybrid between parents from two different genera, a name can be constructed from the two generic names, in part or in entirety (but not both in their entirety) as a condensed formula; × *Mahoberberis* is the name for hybrids between the genera *Mahonia* and *Berberis* (in this case the cross is only bigeneric when *Mahonia*, a name conserved against *Berberis*, is treated as a distinct genus) and × *Fatshedera* is the name for hybrids between the genera *Fatsia* and *Hedera*. The orchid hybrid between *Gastrochilus bellinus*

(Rchb.f.) O.Ktze. and *Doritis pulcherrima* Lindl. carries the hybrid genus name ×*Gastritis* (it has a cultivar called 'Rumbling Tum'!). Alternatively a formula can be used in which the names of the genera are linked by the sign for hybridity '×': *Mahonia* × *Berberis* and *Fatsia* × *Hedera*. Hybrids between parents from three genera are also named either by a formula or by a condensed formula and, in all cases, the condensed formula is treated as a generic name if it is published with a statement of parentage. When published, it becomes the correct generic name for any hybrids between species of the named parental genera. A third alternative is to construct a commemorative name in honour of a notable person and to end it with the termination *-ara*: × *Sanderara* is the name applied to the orchid hybrids between the genera *Brassia*, *Cochlioda* and *Odontoglossum* and commemorates H.F.C. Sander, the British orchidologist.

A name formulated to define a hybrid between two particular species from different genera can take the form of a species name, and then applies to all hybrids produced subsequently from those parent species: × *Fatshedera lizei* Guillaumin is the name first given to the hybrid between *Fatsia japonica* (Thunb.) Decne. & Planch. and *Hedera helix* L. cv. Hibernica, but which must include all hybrids between *F. japonica* and *H. helix*; and × *Cupressocyparis leylandii* (Jackson & Dallimore) Dallimore is the name for hybrids between *Chamaecyparis nootkatensis* (D.Don) Spach and *Cupressus macrocarpa* Hartweg ex Godron. Other examples include × *Achicodonia*, × *Achimenantha*, × *Amarygia*, × *Celsioverbascum*, × *Citrofortunella*, × *Chionoscilla*, × *Cooperanthes*, × *Halimocistus*, × *Ledodendron*, × *Leucoraoulia*, × *Lycene*, × *Osmarea*, × *Stravinia*, × *Smithicodonia*, × *Solidaster* and × *Venidioarctotis*. Because the parents themselves are variable, the progeny of repeated crosses may be distinctive and warrant naming. They may be named under the Botanical Code (prior to 1982 they would have been referred to as *nothomorphs* or bastard forms) and also under the International Code of Nomenclature for Cultivated Plants as 'cultivars': thus, × *Cupressocyparis leylandii* cv. Naylor's Blue. The hybrid nature of × *Sanderara* is expressed by classifying it as a '*nothogenus*' (bastard genus or, in the special circumstances of orchid nomenclature, grex class) and of ×*Cupressocyparis leylandii* by classifying it as a '*nothospecies*' (within a *nothogenus*). For infraspecific ranks the multiplication sign is not used but the term denoting their rank receives the prefix notho-, or 'n-' (*Mentha* × *piperita* L. *nothosubspecies pyramidalis* (Ten.) Harley which, as stated earlier, also implies the autonymous *Mentha* × *piperita nothosubspecies piperita*.

Hybrids between species in the same genus are also named by a formula or by a new distinctive epithet: *Digitalis lutea* L. × *D. purpurea* L. and *Nepeta* × *faassenii* Bergmans ex Stearn are both correct designations for hybrids. In the example of *Digitalis*, the order in which the parents are presented happens to be the correct order, with the seed parent first. It is permissible to indicate the roles of the parents by including the symbols for female '♀' and male '♂', when this information is known, or otherwise to present the parents in alphabetical order.

The orchid family presents particularly complex problems of nomenclature, requiring its own 'Code' in the form of the *Handbook on Orchid Nomenclature and Registration* (Greatwood, Hunt, Cribb & Stewart, 1993). There are some 20,000 species of orchids and to this has been added a huge range of hybrids, some with eight genera contributing to their parentage, and over 70,000 hybrid swarms, or *greges* (singular *grex*–a crowd or troupe), with a highly complex ancestral history.

In cases where a hybrid is sterile because the two sets of chromosomes which it has inherited, one from each parent, are sufficiently dissimilar to cause breakdown of the mechanism which ends in the production of gametes, doubling its chromosome complement may produce a new state of sexual fertility and what is, in effect, a new biological species. Many naturally occurring species are thought to have evolved by such changes and man has created others artificially via the same route, some intentionally and some unintentionally from the wild. The bread-wheats, *Triticum aestivum* L.

are an example of the latter. They are not known in the wild and provide an example of a complex hybrid ancestry but whose name does not need to be designated as hybrid. Even artificially created tetraploids (having, as above, four instead of the normal two sets of chromosomes) need not be designated as hybrid, by inclusion of '×' in the name: *Digitalis mertonensis* Buxton & Darlington is the tetraploid from an infertile hybrid between *D. grandiflora* L. and *D. purpurea* L.

Synonymy and illegitimacy

Inevitably, most plants have been known by two or more names in the past. Since a plant can have only one correct name, which is determined by priority, its other validly published names are synonyms. A synonym may be one which is strictly referable to the same type (a nomenclatural synonym) or one which is referable to another type which is, however, considered to be part of the same taxon (this is a taxonomic synonym). The synonymy for any plant or group of plants is important because it provides a reference list to the history of the classification and descriptive literature on that plant or group of plants.

In the search for the correct name, by priority, there may be names which have to be excluded from consideration because they are regarded as being illegitimate, or not in accordance with the rules.

Names which have the same spelling but are based on different types from that which has priority are illegitimate 'junior homonyms'. Clearly, this prevents the same name being used for different plants. Curiously, this exclusion also applies to the names of those animals which were once regarded as plants, but not to any other animal names.

Published names of taxa, which are found to include the type of an existing name, are illegitimate because they are 'superfluous'. This prevents unnecessary and unacceptable proliferation of names of no real value.

Names of species in which the epithet exactly repeats the generic name have to be rejected as illegitimate 'tautonyms'. It is interesting to note that there are many plant names which have achieved some pleonastic repetition by using generic names with Greek derivation and epithets with Latin derivation: *Arctostaphylos uva-ursi* (bearberry, berry of the bear); *Myristica fragrans* (smelling of myrrh, fragrant), *Orobanche rapum-genistae* (legume strangler, rape of broom); or the reverse of this, *Liquidambar styraciflua* (liquid amber, flowing with storax); *Silaum silaus*; but modern practice is to avoid such constructions. In zoological nomenclature tautonyms are commonplace.

The Code provides a way of reducing unwelcome disturbance to customary usage which would be caused by rigid application of the rule of priority to replace with correct names certain names of families and genera which, although incorrect or problematic are, for various reasons (usually their long usage and wide currency in important literature) agreed to be conserved at a Botanical Congress. These conserved names can be found listed in an Appendix to the Code, together with names which are to be rejected because they are taxonomic synonyms used in a sense which does not include the type of the name, or are earlier nomenclatural synonyms based on the same type, or are homonyms or orthographic variants.

The Code also recommends the ways in which names should be spelt or transliterated into Latin form in order to avoid what it refers to as 'orthographic variants'. The variety found amongst botanical names includes differences in spelling which are, however, correct because their authors chose the spellings when they published them and differences which are not correct because they contain any of a range of defects which have become specified in the Code. This is a problem area in horticultural literature, where such variants are commonplace. It is clearly desirable that a plant name should have a single, constant and correct spelling but this has not been achieved in all fields and reaches its worst condition in the labelling of plants for sale in some nurseries.

The International Code of Nomenclature for Cultivated Plants

There can be no doubt that the diverse approaches to naming garden plants, by common names, by botanical names, by mixtures of botanical and common names, by group names and by fancy names, is no less complex than the former unregulated use of common or vernacular names. The psychology of advertising takes descriptive naming into yet new dimensions. It catches the eye with bargain offers of colourful, vigorous and hardy, large-headed, incurved *Chrysanthemum* cvs. by referring to them as HARDY FOOTBALL MUMS. However, we are not here concerned with such colloquial names or the ethics of mail-order selling techniques but with the regulation of meaningful names under the Code.

In 1952, the Committee for the Nomenclature of Cultivated Plants of the International Botanical Congress and the International Horticultural Congress in London adopted the International Code of Nomenclature for Cultivated Plants. Sometimes known as the Cultivated Code, it was first published in 1953 and has been revised several times at irregular intervals since then (Trehane, 1995). This Code formally introduced the term 'cultivar' to encompass all varieties or derivatives of wild plants which are raised under cultivation and its aim is to 'promote uniformity and fixity in the naming of agricultural, sylvicultural and horticultural cultivars (varieties)'. The term *culton* (plural *culta*) is also mooted as an equivalent of the botanical term *taxon*.

The Cultivated Code governs the names of all plants which retain their distinctive characters, or combination of distinctive characters, when reproduced sexually (by seed), or vegetatively in cultivation. Because the Code does not have legal status, the commercial interests of plant breeders are guarded by the Council of the International

Union for the protection of New Varieties of Plants (UPOV). In Britain, the Plant Varieties Rights Office works with the Government to have UPOV's guidelines implemented. Also, in contrast with the International Code for Botanical Nomenclature, the Cultivated Code faces competition from legislative restraints presented by commercial law in certain countries. Where national and international legislation recognize 'variety' as a legal term and also permit commercial trade designation of plant names, such legislative requirements take precedence over the Rules of the Cultivated Code.

The Cultivated Code accepts the International Rules of Botanical Nomenclature and the retention of the botanical names of those plants which are taken into cultivation from the wild and has adopted the same starting date for priority (precedence) of publication of cultivar names (*Species Plantarum* of 1753). It recognizes only the one category of garden-maintained variant, the cultivar (cv.) or garden variety, which should not be confused with the botanical *varietas*. It recognizes also the supplementary, collective category of the Cultivar Group, intermediate between species and cultivar, for special circumstances explained below. The name of the Cultivar Group is for information and may follow the cultivarietal name, being placed in parentheses: *Solanum tuberosum* 'Desiré' (Maincrop Group) or potato 'Desiré' (Maincrop Group).

Unlike wild plants, cultivated plants are maintained by unnatural treatment and selection pressures by man. A cultivar must have one or more distinctive attributes which separate it from its relatives and may be:

1 Clones derived asexually from (a) a particular part of a plant, such as a lateral branch to give procumbent offspring, (b) a particular phase of a plant's growth cycle, as from plants with distinctive juvenile and adult phases, (c) an aberrant growth, such as a gall or witches' broom.
2 Graft chimaeras (which are dealt with below).

3 Plants grown from seed resulting from open pollination, provided that their characteristic attributes remain distinctive.
4 Inbred lines resulting from repeated self-fertilization.
5 Multilines, which are closely related inbred lines with the same characteristic attributes.
6 F_1 hybrids, which are assemblages of individuals that are re-synthesized only by crossbreeding.
7 Topovariants, which are repeatedly collected from a specific provenance (equivalent to botanical ecospecies or ecotypes).
8 Assemblages of genetically modified plants.

The cultivar's characteristics determine the application of the name – so genetic diversity may be high and the origins of a single cultivar may be many. If the method of propagating the cultivar is changed and the offspring show new characteristics, they may not be given the name of the parent cultivar. If any of the progeny revert to the parental characteristics, they may carry the parental cultivar name.

Plants grafted onto distinctive rootstocks, such as apples grafted onto Malling dwarfing rootstocks, may be modified as a consequence but it is the scion which determines the cultivar name – not the stock. Plants which have their physical form maintained by cultural techniques, such as bonsai and topiary subjects and fruit trees trained as espaliers, etc. do not qualify for separate cultivar naming since their characteristics would be lost or changed by cessation of pruning or by pruning under a new regime.

From this it will be seen that with the single category of cultivar, the hybrid between parents of species rank or any other rank has equal status with a 'line' selected within a species, or taxon of any other rank, including another cultivar, and that parity exists only between names, not between biological entities. The creation of a cultivar name does not, therefore, reflect a fragmentation of the parent taxon but does reflect the existence of a group of plants having a particular set of features, without definitive reference to its parents. Features may be concerned with cropping, disease resistance or biochemistry, showing that the Cultivated Code requires a greater flexibility than the Botanical Code. It achieves this by having no limiting requirement for 'typical' cultivars but by regarding cultivars as part of an open system of nomenclature. Clearly, this permits a wide range of applications and differences with the Botanical Code and these are considered in Styles (1986).

The names of cultivars have had to be 'fancy names' in common language and not in Latin. Fancy names come from any source. They can commemorate anyone, not only persons connected with botany or plants, or they can identify the nursery of their origin, or be descriptive, or be truly fanciful. Those which had Latin garden-variety names were allowed to remain in use: *Nigella damascena* L. has two old varietal names *alba* and *flore pleno* and also has a modern cultivar with the fancy name cv. Miss Jekyll. In the glossary, no attempt has been made to include fancy names but a few of the earlier Latin ones have been included.

In order to be distinguishable, the cultivar names have to be printed in a typeface unlike that of the species name and to be given capital initials. They also have to be either preceded by 'cv.', as above, or placed between single quotation marks. Thus, *Salix caprea* L. cv. Kilmarnock, or *S. caprea* 'Kilmarnock', is a weeping variety of the goat willow and is also part of the older variety *Salix caprea* var. *pendula*. Other examples are *Geranium ibericum* Cav. cv. *Album* and *Acer davidii* Franchet 'George Forrest'. The misuse of the apostrophe that is now commonplace may require the use of single quotation marks to be changed in the future.

Cultivar names can be attached to an unambiguous common name, such as potato 'Duke of York' for *Solanum tuberosum* L. cv. Duke of York, or to a generic name such as *Cucurbita* 'Table Queen' for *Cucurbita pepo* L. cv. Table Queen, or of course to the botanical name, even when this is below the rank of species; *Rosa sericea* var. *omeiensis* 'Praecox'.

Commercial breeders have produced enormous numbers of cultivars and cultivar names. Some have found popularity and have therefore persisted and remained available to gardeners but huge numbers have not done so and have been lost or remain only as references in the literature. The popular practice of naming new cultivars for people (friends, growers, popular personalities or royalty) or the nursery originating the new cultivar is a form of flattery. For those honouring people who made some mark upon horticulture during their lifetime it is more likely that we can discover more about the plant bearing their name but, for the vast majority of those disappearing into obscurity, the only record may be the use of their name in a nurseryman's catalogue. Alex Pankhurst (1992) has compiled an interesting collection of commemorative cultivarietal names.

For some extensively bred crops and decorative plants there is a long-standing supplementary category, the Cultivar Group. By naming the Cultivar Group in such plants, a greater degree of accuracy is given to the garden name; such as pea 'Laxton's Progress' (Wrinkle-seeded Group), and *Rosa* 'Albéric Barbier' (Rambler Group) and *Rosa* 'Agnes'(Rugosa Group). However, for some trade purposes a cultivar may be allocated to more than one Cultivar Group ; such as potato 'Desiré' (Maincrop Group) but also potato 'Desiré' (Red-skinned Group).

The same cultivar name may not be used twice within a genus, or denomination class, if such duplication would cause ambiguity. Thus, we could never refer to cherries and plums by the generic name, *Prunus*, alone. Consequently, the same fancy name could not be used for a cultivar of a cherry and for a cultivar of a plum. Thus, the former cultivars Cherry 'Early Rivers' and Plum 'Early Rivers' are now Cherry 'Early Rivers' and Plum 'Rivers Early Prolific'.

To ensure that a cultivar has only one correct name, the Cultivated Code requires that priority acts and, to achieve this, publication and registration are necessary. To establish a cultivar name, publication has to be in printed matter which is dated and distributed to the public. For the more popular groups of plants, usually genera, there are societies which maintain statutory registers of names and the plant breeding industry has available to it the Plant Variety Rights Office as a statutory registration body for crop-plant names as trade marks for commercial protection, including patent rights on vegetatively propagated cultivars. Guidance on all these matters is provided as appendices to the Code.

As with botanical names, cultivars can have synonyms. However, it is not permissible to translate the fancy names into other languages using the same alphabet; except that in commerce the name can be translated and used as a trade designation. This produces the confusion that, for example, *Hibiscus syriacus* 'Blue Bird' is just a trade name for *Hibiscus syriacus* 'L'Oiseau Bleu' but will be the one presented at the point of sale. Also, translation is permitted to or from another script and the Code provides guidance for this.

In the case of the names of Cultivar Groups, translation is permitted; since these are of the nature of descriptions that may relate to cultivation. An example provided is the Purple-leaved Group of the beech which is the Purpurblätterige Gruppe in German, the Gruppo con Foglie Purpuree in Italian and the Groupe à Feuilles Pourpres in French.

For the registration of a new cultivar name, it is also recommended that designated standards are established. These may be herbarium specimens deposited in herbaria, or illustrations that can better define colour characteristics, or documentation held at a Patents Office or a Plant Variety Protection Office. In each case, the intention is that they can be used as reference material in determining later proposed names. This brings the Cultivated Code closer to the Botanical Code and is a small step towards the eventual establishment of an all-encompassing Code of Bionomenclature.

When the names of subspecies, varieties and forms are used, it is a growing trend to present the full name without indication of these – particularly in America, but also in our own horticultural literature (Bagust, 2001), as a shorthand cross reference. Thus,

Narcissus bulbocodium subsp. *bulbocodium* var. *conspicuus* is written as *Narcissus bulbocodium bulbocodium conspicuus*. This is confusing when the cultivar name has a Latin form since this then appears as a pre-Linnaean phrase name (e.g. *Narcissus albus plenus odoratus* and *Rosa sericea omiensis praecox*).

Graft chimaeras

One group of plants which is entirely within the province of gardening and the Cultivated Code is that of the graft chimaeras, or graft hybrids. These are plants in which a mosaic of tissues from the two parents in a grafting partnership results in an individual plant upon which shoots resembling each of the parents, and in some cases shoots of intermediate character, are produced in an unpredictable manner. Unlike sexually produced hybrids, the admixture of the two parents' contributions is not at the level of the nucleus in each and every cell but is more like a marbling of a ground tissue of one parent with streaks of tissue of the other parent. Chimaeras can also result from mutation in a growing point, from which organs are formed composed of normal and mutant tissues, as with genetic forms of variegation. In all cases, three categories may be recognized in terms of the extent of tissue 'marbling', called sectorial, mericlinal and periclinal chimaeras. The chimaeral condition is denoted by the addition sign '+' instead of the multiplication sign '×' used for true hybrids. A chimaera which is still fairly common in Britain is that named +*Laburnocytisus adamii* C.K.Schneider. This was the result of a graft between *Cytisus purpureus* Scop. and *Cytisus laburnum* L., which are now known as *Chamaecytisus purpureus* (Scop.) Link and *Laburnum anagyroides* Medicus, respectively. Although its former name *Cytisus* + *adamii* would not now be correct, the name *Laburnocytisus* meets the requirement of combining substantial parts of the two parental generic names, and can stand.

Combining generic names for Graft chimaeras must not duplicate a composite name for a sexually produced hybrid between the same progenitors. Hybrids between species of *Crataegus* and species of *Mespilus* are ×*Crataemespilus* but the chimaera between the same species of the same genera is +*Crataegomespilus*. As in this example, the same progenitors may yield distinctive chimaeras and these may be given cultivar names: + *Crataegomespilus* 'Dardarii' and +*Crataegomespilus* 'Jules d'Asnières'.

It is interesting to speculate that if cell- and callus-culture techniques could be used to produce chimaeral mixtures to order, it may be possible to create some of the conditions which were to have brought about the early 'green revolutions' of the 1950–2000 period. Protoplast fusion methods failed to combine the culturally and economically desirable features of distant parents, which were to have given multi-crop plants and new nitrogen-fixing plants, because of the irregularities in fusion of both protoplasts and their nuclei. It may be that intact cells would prove easier to admix. However, molecular genetics and genetic manipulation have shown that genetic control systems can be modified in ways which suggest that any aspect of a plant can, potentially, be manipulated to suit man's requirements and novel genetic traits can be inserted into a plant's genome by using DNA implants. The genetically modified (GM) results of such manipulation are the products of commercial undertakings and may be given cultivar names but are protected commercially by trade designations.

Botanical terminology

There is nothing accidental about the fact that in our everyday lives we communicate at two distinct levels. Our 'ordinary' conversation employs a rich, dynamic language in which meaning can differ from one locality to another and change from time to time. Our 'ordinary' reading is of a written language of enormous diversity – ranging from contemporary magazines which are intentionally erosive of good standards, to high-quality prose of serious writers. However, when communication relates to specific topics, in which ambiguity is an anathema, the language which we adopt is one in which 'terminology' is relied upon to convey information accurately and incontrovertibly. Thus, legal, medical and all scientific communications employ terms which have widely accepted meanings and which therefore convey those meanings in the most direct way. Because, like botanical terms for the parts of the flower, these terms are derived predominantly from classical roots and have long-standing acceptance, they have the added advantage of international currency.

This glossary contains many examples of words which are part of botanical terminology as well as being employed as descriptive elements of plant names. Such is the wealth of this terminology that an attempt here to discriminate between and explain all the terms which relate, say, to the surface of plant leaves and the structures (hairs, glands and deposits) which subscribe to that texture would make tedious reading. However, terms which refer to such conspicuous attributes as leaf shape and the form of inflorescences are very commonly used in plant names and, since unambiguous definition would be lengthy, are illustrated as figures.

More extensive glossaries of terminology can be found in textbooks and floras but the 6th edition (1955) of Willis's *Dictionary of*

Flowering Plants and Ferns (1931) is a particularly rewarding source of information, and B.D. Jackson's *Glossary of Botanic Terms* (1960) is a first rate source of etymological information.

The glossary

The glossary is for use in finding the meanings of the names of plants. There are many plant names which cannot be interpreted or which yield very uninformative translations. Authors have not always used specific epithets with a single, narrow meaning so that there is a degree of latitude in the translation of many epithets. Equally, the spelling of epithets has not remained constant, for example in the case of geographic names. The variants, from one species to another, are all correct if they were published in accordance with the Code. In certain groups such as garden plants from, say, China and exotics such as many members of the profuse orchid family, commemorative names have been applied to plants more frequently than in other groups. If the reader wishes to add further significance to such names, he will find it mostly in literature on plant-hunting and hybridization or from reference works such as that on taxonomic literature by Stafleu & Cowan (1976–).

Generic names in the European flora are mostly of ancient origin. Their meanings, even of those which are not taken from mythological sources, are seldom clear and many have had their applications changed and are now used as specific epithets. Generic names of plants discovered throughout the world in recent times have mostly been constructed to be descriptive and will yield to translation. The glossary contains the generic names of a wide range of both garden and wild plants and treats them as singular nouns, with capital initials. Orthographic variants have not been sought out but a few are presented and have the version which is generally incorrect between parentheses. Listings of generic names can be found in Farr (1979–86) and in Brummitt (1992).

As an example of how the glossary can be used, we can consider the name *Sarcococca ruscifolia*. This is the name given by Stapf to plants which belong to Lindley's genus *Sarcococca*, of the family *Buxaceae*, the box family. In the glossary we find *sarc-*, *sarco-* meaning fleshy and *-coccus -a -um* meaning 'berried' and from this we conclude that *Sarcococca* means Fleshy-berry (the generic name being a singular noun) and has the feminine gender. We also find *rusci-* meaning butcher's-broom-like or resembling *Ruscus* and *-folius -a -um* meaning -leaved and we conclude that this species of Fleshy-berry has leaves resembling the prickly cladodes (leaf-like branches) of *Ruscus*. The significance of this generic name lies in the fact that dry fruits are more typical in members of the box family than fleshy ones.

From this example, we see that names can be constructed from adjectives or adjectival nouns to which prefixes or suffixes can be added, thus giving them further qualification. As a general rule, epithets which are formed in this way have an acceptable interpretation when '-ed' is added to the English translation; this would render *ruscifolia* as '*Ruscus*-leaved'.

It will be noted that *Sarcococca* has a feminine ending (*-a*) and that *ruscifolia* takes the same gender. However, if the generic name had been of the masculine gender the epithet would have become *ruscifolius* and if of the neuter gender then it would have become *ruscifolium*. For this reason the entries in the glossary are given all three endings which, as pointed out earlier, mostly take the form *-us -a -um* or *-is -is -e*.

Where there is the possibility that a prefix which is listed could lead to the incorrect translation of some epithet, the epithet in question is listed close to the prefix and to an example of an epithet in which the prefix is employed. Examples are:

aer- meaning air- or mist-, gives *aerius -a -um* meaning airy or lofty; *aeratus -a -um*, however, means bronzed (classically, made of bronze).

caeno-, from the Greek χαινος, means fresh-, but
caenosus -a -um, is from the Latin *caenum* and means muddy or growing on mud or filth.

Examples will be found of words which have several fairly disparate meanings. A few happen to reflect differences in meaning of closely similar Greek and Latin source words as in the example above and others reflect what is to be found in literature, in which other authors have suggested meanings of their own. Similarly, variations in spelling are given for some names and these are also to be found in the literature although not all of them are strictly permissible for nomenclatural purposes. Their inclusion emphasizes the need for uniformity in the ways in which names are constructed and provides a small warning that there are in print many deviant names, some intentional and some accidental.

Many of the epithets which may cause confusion are either classical geographic names or terms which retain a meaning closer to that of the classical languages. There are many more such epithets than are listed in this glossary.

The glossary

a-, *ab-* away from-, downwards-, without-, un-, very-
abbreviatus -a -um shortened
Abelia for Dr Clarke Abel (1780–1826), physician and writer on China
Abeliophyllum *Abelia*-leaved (similarity of foliage)
aberconwayi for Lord Aberconway, former President of the RHS
Abies Rising-one (the ancient Latin name for a tall tree)
abietinus -a -um *Abies*-like, fir-tree-like
-abilis -is -e -able, -capable of (preceded by some action)
abnormis -is -e departing from normal in some structure
abortivus -a -um with missing or malformed parts
abros delicate
abrotani-, *abrotonoides* *Artemisia*-like (from an ancient Greek name, αβροτονον, for wormwood or mugwort)
abrotanum ancient Latin name for southernwood
abruptus -a -um ending suddenly, blunt-ended
abscissus -a -um cut off
absinthius -a -um from an ancient Greek or Syrian name for wormwood
Absynthium the old generic name, αψινθιον, for wormwood
Abutilon the Arabic name for a mallow
abyssinicus -a -um of Abyssinia, Abyssinian
ac-, *ad-*, *af-*, *ag-*, *al-*, *an-*, *ap-*, *ar-*, *as-*, *at-* near-, towards-
Acacia Thorn (from the Greek ακις)
Acaena Thorny-one (ακη, ακις)
Acalypha Nettle-like (from the Greek name, ακελπε, for a nettle) the hispid leaves
acantho-, *acanthus* thorny-, spiny-

Acantholimon Thorny-*Limonium*

Acanthopanax Spiny-*Panax* (the prickly nature of the plants)

Acanthus Prickly-one, the Nymph, ακανθα, loved by Apollo was changed into an *Acanthus*

acaulis -is -e, acaulos -os -on lacking an obvious stem

accicus -a -um with a small acute apical cleft, emarginate

Acer Sharp (Ovid's name for a maple, either from its use for lances or its leaf-shape), etymologically linked to oak and acre

acer, acris, acre sharp-tasted, acid

Aceras Without-a-horn (the lip has no spur, α–κερας)

acerbus -a -um harsh-tasted

aceroides maple-like

acerosus -a -um pointed, needle-like

acetabulosus -a -um saucer-shaped

acetosus -a -um slightly acid, sour

acetosellus -a -um slightly acid

-aceus -a -um -resembling (preceded by a plant name), (rose)aceous

Achillea after the Greek warrior Achilles, αχιλλευς, reputed to have used it to staunch wounds

Achimenes Tender-one (χειμινο cold-hating)

Achras an old Greek name, αχρας, for a wild pear used by Linnaeus for the sapodilla or chicle tree

Achyranthes Chaff-flower, αχυρον–ανθος

acicularis -is -e needle-shape

aciculatus -a -um finely marked as with needle scratches

aciculus -a -um sharply pointed (e.g. leaf-tips)

acidosus -a -um acid, sharp, sour

acidotus -a -um sharp-spined

acinaceus -a -um, aciniformis -is -e scimitar-shaped

acinifolius -a -um *Acinos*-leaved, basil-thyme-leaved

Acinos Dioscorides' name for a heavily scented calamint

Acioa Pointed (the toothed bracts of some species)

acmo- pointed- (followed by a part of a plant), anvil-shaped-

Acokanthera Pointed-anther (ακοκε a point)

aconiti- aconite-

Aconitum the name of a hill in Ponticum, used by Theophrastus for poisonous aconite

Acorus Without-pupil, Dioscorides' name, ακορον, for an iris (its use in treating cataract)

acr-, acro- summit-, highest-, ακρα (followed by noun e.g. hair, or verb e.g. fruiting)

Acradenia Apical-gland

acreus -a -um of high places

Acrobolbus Apical-bulb (the archegonia are surrounded by minute leaves at the apex of the stem)

Acroceras Apex-horn (the glumes have an excurrent vein at the tip)

Acrostichum Upper-spotted, ακρα–στικτος (the sori cover the backs or whole of the upper pinnae)

Actaea from the Greek name, ακτεα, for elder (the shape of the leaves)

actin-, actino- radiating-, ακτις–ινος (followed by a part of a plant)

actinacanthus -a -um ray-spined

actinia sea-anemone

Actinidia Rayed (ακτινος a ray) (refers to the radiate styles)

actinius -a -um sea-anemone-like

Actinocarpus Radiate-fruit (the spreading ripe carpels of thrumwort)

acu- pointed-, acute-

aculeatus -a -um having prickles, prickly, thorny

aculeolatus -a -um having small prickles or thorns

aculiosus -a -um decidedly prickly

acuminatus -a -um with a long, narrow and pointed tip (see Fig. 7(*c*)), acuminate

acuminosus -a -um with a conspicuous long flat pointed apex

acutus -a -um, acuti- acutely pointed, sharply angled at the top

adamantinus -a -um from Diamond Lake, Oregon, USA

Adansonia for Michel Adanson (1727–1806), French West African botanist, baobab

aden-, adeno- gland-, glandular-, αδμν

Adenocarpus Gland-fruit (the glandular pod)
Adenophora Gland-bearing
adenophyllus -a -um glandular-leaved
adenotrichus -a -um glandular-hairy
Adiantum Unwettable, the old Greek name, αδιαντον, refers to its staying unwetted under water
Adlumia for Maj. John Adlum (1759–1836), American vitculturist
admirabilis -is -e to be admired, admirable
adnatus -a -um attached through the whole length, adnate (e.g. anthers)
Adonis for the Greek god loved by Venus, killed by a boar, from whose blood grew a flower; an *Anemone*
Adoxa Without-glory, α–δοξα (its small greenish flowers)
adpressus -a -um pressed together, lying flat against (e.g. the hairs on the stem)
adriaticus -a -um from the Adriatic region
Adromiscus Stout-stemmed (αδρος sturdy)
adscendens curving up from a prostrate base, half-erect, ascending
adstringens constricted
adsurgens rising up, arising
adulterinus -a -um of adultery (intermediate between two other species suggesting hybridity, as in *Asplenium adulterinum*)
aduncus -a -um hooked, having hooks
adustus -a -um fuliginous, soot-coloured
advena exotic, stranger
Aechmea Pointed (αιχημ a point)
aegaeus -a -um of the Aegean region
Aegopodium Goat's-foot, αιγος–ποδιον (the leaf shape)
aelophilous -a -um wind-loving (plants disseminated by wind)
aemulus -a -um imitating, rivalling
aeneus -a -um bronzed
Aeonium the Latin name from the Greek, αιον, for age
aequi-, aequalis -is -e, aequali- equal-, equally-
aequilateralis -is -e, aequilaterus -a -um equal-sided

aequinoctialis -is -e   of the equinox (the flowering time)
aer-   air-, mist-
aeranthos -os -on   air-flower (not ground-rooted)
Aeranthus   Air-flower (rootless epiphyte)
aeratus -a -um   bronzed
aereus -a -um   copper-coloured
Aerides   Of the air (epiphytic)
aerius -a -um   airy, lofty
aeruginosus -a -um   rusty, verdigris-coloured
Aeschynanthus   Shame-flower (αισχος shame)
aeschyno-   shy-, to be ashamed-
aesculi-   horse-chestnut-like
Aesculus   Linnaeus' name from the Roman name of an edible acorn. The Turks reputedly used 'conkers' in a treatment for bruising in horses – now attributed to its aescin content
aestivalis -is -e   of summer
aestivus -a -um   developing in the summer
aestuans   glowing
aestuarius -a -um   of tidal estuaries
aethereus -a -um   aerial (epiphytic)
Aethionema   Strange-filaments (those of the long stamens are winged and toothed)
aethiopicus -a -um   of Africa, African, of NE Africa
Aethusa   Burning-one, αιθω (its pungency)
aetiolatus -a -um   lank and yellowish, etiolated
aetnensis -is -e   from Mt Etna, Sicily
aetolicus -a -um   from Aetolia, Greece
-aeus   -belonging to (of a place)
afer, afra, afrum   Afro-, African
affinis -is -e   related, similar to
aflatunensis -is -e   from Aflatun, central Asia
africanus -a -um   African
Afzelia   for Adam Afzelius (1750–1837), Swedish botanist
Agapanthus   Good-flower (αγαπη good)

Agapetes Desirable
Agathis Ball-of-twine (the appearance of the cones)
Agastache Much-spiked
agastus -a -um charming, pleasing
Agathelpis Good-hope (its natural area on the Cape)
Agave Admired-one (αγαυε noble)
Ageratum Un-ageing (the flower-heads long retain their colour), from a name, αγηρατος, used by Dioscorides
agetus -a -um wonderful
agglomeratus -a -um in a close head, congregated together
agglutinatus -a -um glued or firmly joined together
aggregatus -a -um clustered together
aglao- bright- (αγλαος)
Aglaodorum Bright-bag (the spathe around the inflorescence)
Aglaonema Bright-thread (possibly the naked male inflorescences)
agninus lamb
agnus-castus chaste-lamb
-ago -like
agrarius -a -um, *agrestis -is -e* of fields, wild on arable land
agri-, *agro-* grassy-, grass-like-, field-, meadow-
agricola of the fields, rustic
Agrimonia Cataract (from its medicinal use), Pliny's transliteration of *argemonia*
Agropyron(um) Field-wheat
Agrostemma Field-garland (Linnaeus' view of its suitability for such)
Agrostis Field (-grass)
ai-, *aio-* eternally-, always-
Ailanthus Reaching-to-heaven (from a Moluccan name)
aiophyllus -a -um always in leaf, evergreen
Aira an old Greek name, αιρα, for darnel grass
Aitonia for William Aiton (1759–1793), head gardener at Kew
aizoides resmbling *Aizoon*
Aizoon Always-alive, αει–ζωη

ajacis -is -e   of Ajax, son of Telemon, from whose blood grew a hyacinth marked AIA

ajanensis -is -e   from Ajan, E Asia

Ajania   from Ajan, E Asia (*Chrysanthemum*)

Ajuga   corrupted Latin for abortifacient (not-yoked)

ajugi-   *Ajuga-*, bugle-

Akebia   the Japanese name, akebi

alabastrinus -a -um   like alabaster

Alangium   from an Adansonian name for an Angolan tree

alaris -is -e   winged, alar (axillary)

alaternus   an old generic name for a buckthorn

alatus -a -um, alati-, alato-   wing-like (fruits), winged (stems with protruding ridges which are wider than thick), alate

alb-, albi-, albo-, -albus -a -um   white-, -white

albatus -a -um   turning white

albens   white

Alberta   for Albertus Magnus (1193–1280) (*A. magna* is from Natal)

albertii, albertianus -a -um   for Albert, Prince Consort, or for Dr Albert Regel, Russian plant collector in Turkestan

albescens   turning white

albicans, albidus -a -um, albido-, albulus -a -um   whitish

Albizia (Albizzia)   for Filippo degli Albizzi, Italian naturalist

Albuca   White

albus -a -um, albi-, albo-   dead-white

Alcea   the name, αλκαια, used by Dioscorides

alceus -a -um   mallow-like, resembling *Alcea*

Alchemilla   from Arabic, alkemelych, reference either to its reputed magical properties or to the fringed leaves of some species

alcicornis -is -e   elk-horned

aleppicus -a -um   of Aleppo, N Syria

aleur-, aleuro-   mealy-, flowery-, αλευρον (surface texture)

aleuticus -a -um   Aleutian

algidus -a -um   cold, of high mountains

Alhagi   the Mauritanian name

alicae for Princess Alice of Hesse (1843–1878)
alicia for Miss Alice Pegler, plant collector in Transkei, South Africa
-alis -is -e -belonging to (a place)
Alisma Dioscorides' name, αλισμα, for a plantain-leaved water plant
Alkanna from the Arabic, al-henna, for *Lawsonia inermis*, henna. Hence our 'alkanet'
alkekengi a name, αλκικακαβον, used by Dioscorides
Allamanda for Dr Frederick Allamand, who sent seeds of this to Linnaeus, from Brazil
allantoides sausage-shaped
allatus -a -um introduced, not native, foreign
Allexis Different (distinct from *Rinorea*)
alliaceus -a -um, *allioides* *Allium*-like, smelling of garlic
Alliaria Garlic-smelling
allionii for Carlo Allioni (1705–1804), author of *Flora Pedemontana*
Allium the ancient Latin name for garlic
allo- diverse-, several-, different-, other- (αλλος)
Allosorus Variable-sori, αλλος–σωρος (their shapes vary)
almus -a -um bountiful
alni- *Alnus*-like-, alder-like-
Alnus the ancient Latin name for the alder
Alocasia from the Greek, καλοκασια, attractive (leaves)
Aloe from the Semitic, alloeh, for the medicinal properties of the dried juice (αλοη) (Aloë, of Linnaeus)
aloides *Aloë*-like
Alonsoa for Alonzo Zanoni, Spanish official in Bogotá
Alopecurus Fox-tail, αλωπηξ–ουρα
Aloysia for Queen Maria Louisa of Spain
alpester -ris -re of mountains, of the lower Alps
alpicolus -a -um of high mountains
Alpinia for Prosper Alpino (1553–1616), Italian botanist
alpinus -a -um of upland or mountainous regions, alpine, of the high Alps

alsaticus -a -um from Alsace, France

Alsine a name, αλσινη, used by Theophrastus for a chickweed-like plant

alsinoides chickweed-like

also- leafy glade-, of groves-

Alsophila, alsophilus -a -um Grove-loving

Alstonia for Prof. Charles Alston (1716–1760), of Edinburgh

alstonii for Capt. E. Alston, collector of succulents in Ceres, South Africa

Alstroemeria for Baron Claus Alstroemer

altaclerensis -is -e from High Clere Nurseries (Alta Clara)

altamahus -a -um from the Altamata River

alternans alternating

Alternanthera Alternating-stamens

alterni-, alternus -a -um alternating on opposite sides, every other-, alternate

Althaea Healer, a name, αλθαια, used by Theophrastus

alti-, alto-, altus -a -um tall, high

altilis -is -e nutritious, fat, large

alumnus -a -um well-nourished, flourishing

alutaceus -a -um of the texture of soft leather, *alutus*

alveolatus -a -um with shallow pits, alveolar

Alyssum Pacifier (an ancient Greek name, αλυσσος, without-fury)

amabilis -is -e pleasing, lovely

Amaranthus (Amarantus) unfading-flower (αμαραντος lasting flower)

amaranticolor purple, *Amaranthus*-coloured

amarellus -a -um, amarus -a -um bitter (as in the amaras or bitters of the drinks industry, e.g. *Quassia amara*)

Amaryllis the name of a country girl in Virgil's writings

amaurus -a -um dark, without lustre

amazonicus -a -um from the Amazon basin, South America

amb-, ambi- around-, both-

ambianensis -is -e   from Amiens, France
ambigens, ambiguus -a -um   doubtful, of uncertain relationship
ambly-   blunt-
amblygonus -a -um   blunt-angled
amblyodon   blunt-toothed
amboinensis -is -e   from Amboina, Indonesia
ambrosia   elixir of life, food of the gods giving immortality, divine food
Amelanchier   a Provençal name for *A. ovalis* (snowy-*Mespilus)*
amelloides   resembling-*Amellus*
Amellus   a name used by Virgil for a blue-flowered composite from the River Mella
amentaceus -a -um   having catkins, catkin-bearing
amenti-   catkin-
americanus -a -um   from the Americas
amesianus -a -um   for Frederick Lothrop Ames, American orchidologist, or for Prof. Oakes Ames of Harvard Botanic Garden and orchidologist
amethystea, amethystinus   the colour of amethyst gems, violet
amicorum   of the Friendly Isles, Tongan
amictus -a -um   clad, clothed
amiculatus -a -um   cloaked, mantled
Ammi   Sand, a name used by Dioscorides
ammo-   sand-
ammoniacum   gum ammoniac, an old generic name
Ammophila   Sand-lover, αμμος–φιλος
ammophillus -a -um   sand-loving (the habitat)
amoenulus -a -um   quite pleasing or pretty
amoenus -a -um   charming, delightful, pleasing
amomum   purifying (the Indian spice plant *Amomum* was used to cure poisoning)
Amorpha   Deformed-one the genus of the greyish-downy lead plant, *Amorpha canescens*, lacks wing and keel petals

Amorphophallus Deformed-mace (the enlarged spadix)
amorphus -a -um, amorpho- deformed (αμορφος), shapeless, without form
ampelo- wine-, vine-, grape-
ampeloprasum leek of the vineyard, a name, αμπελοπρασσυ, in Dioscorides
Ampelopsis Vine-resembling
amphi-, ampho- on-both-sides, in-two-ways-, both-, double-, of-both-kinds-, around-
amphibius -a -um with a double life, growing both on land and in water
Amphorella Small-wine-jar
amplectans stem-clasping (leaf bases)
amplexicaulis -is -e embracing the stem (e.g. the base of the leaf, see Fig. 6(*d*))
amplexifolius -a -um leaf-clasping
ampli- large-, double-
amplissimus -a -um very large, the biggest
amplus -a -um large
ampulli- bottle-
ampullaceus -a -um, ampullaris -is -e bottle-shaped, flask-shaped
Amsinckia for W. Amsinck (1752–1831), of Hamburg
amurensis -is -e from the region of the Amur River, E Siberia
amygdalinus -a -um of almonds, almond-like, kernel-like
amygdalus the Greek name, αμυγδαλινος, for the almond tree
an-, ana- upon-, without-, backwards-, above-, again-, upwards-, up-
Anacamptis Bent-back (the long spur of the flower)
anacanthus -a -um lacking thorns
Anacardium Heart-shaped (Linnaeus' name refers to the shape of the false-fruit)
Anacharis Without-charm
Anagallis Unpretentious, without-boasting, without-adornment
anagyroides resembling *Anagyris*, curved backwards

Ananas probably of Peruvian origin
Anaphalis Greek name for an immortelle, derivation obscure
anastaticus -a -um rising up (*Anastatica hierochuntica* resurrection plant or rose of Jericho)
anatolicus -a -um from Anatolia, Turkish
anatomicus -a -um skeletal (leaves)
anceps two-edged (stems), two-headed, ambiguous
Anchusa Strangler (Aristophanes' name, αγχουσα, formerly for a plant yielding a red dye)
ancistro- fish-hook-
ancylo- hooked-
andegavensis -is -e from Angers in Anjou, France
Andersonia for Wm. Anderson, botanist on Cook's 2nd and 3rd voyages
andersonii for Thos. Anderson, botanist in Bengal or for J. Anderson who collected in the Gold Coast (Ghana)
andicolus -a -um from the central Andean cordillera
Andrachne the ancient Greek name, ανδραχνε, for an evergreen shrub
andro-, andrus -a -um stamened-, anthered-, male, man (ανδρο–)
androgynus -a -um with staminate and pistillate flowers on the same head
Andromeda after the daughter of Cepheus and Cassiope rescued by Perseus from the sea monster
Andropogon Bearded-male (awned male spikelet)
Androsace Man-shield (the exposed stamens of heterostyled spp.)
Androsaemum Man's-blood (the blood-coloured juice of the berries)
Anemia (Aneimia) Naked (ανειμον), the sori have no indusia
Anemone a name used by Theophrastus. Possibly a corruption of Naaman, a Semitic name for Adonis whose blood sprung the crimson-flowered *Anemone coronaria*. Commonly called windflower
anfractuosus -a -um twisted, bent, winding
Angelica from the Latin for an angel (healing powers, see *Archangelica*)
angio- urn-, vessel-, enclosed- (boxed), αγγειον

anglicus -a -um, anglicorum English, of the English
Angraecum a Malayan name, angurek, for epiphytes
angui-, anguinus -a -um serpentine, eel-like-, wavy (*anguilla* a serpent)
angularis -is -e angular
anguligerus -a -um having hooks, hooked
angulosus -a -um having angles, angular
angusti-, angustus -a -um narrow
angustiflius -a -um narrow-leaved
angustior narrower
Anigosanthus Open-flower (ανοιγος)
Anisantha Unequal-flower (flowers vary in their sexuality)
anisatus -a -um aniseed-scented
aniso- unequally-, unequal-, uneven-, ανισος, anise-smelling-
Anisophyllea Unequal-leaved (the paired large maturing and small transient leaves)
Anisum Aniseed (an old generic name)
ankylo- crooked-, αγκυλος
Annona (Anona) from the Haitian name
annotinus -a -um one year old, of last year (with distinct annual increments)
annularis -is -e, annulatus -a -um ring-shaped, having rings (markings)
annuus -a -um annual
ano- upwards-, up-, ανω, towards the top-
Anogeissus Towards-the top-tiled (the scale-like fruiting heads)
Anogramma Towards-the-top-lined (the sori first mature towards the tips of the pinnae)
anomalus -a -um unlike its allies, out of the ordinary
Anonidium Like-*Annona*
anopetalus -a -um erect-petalled
Anopyxis Upright-capsule (the fruit is held upright until it dehisces)
anosmus -a -um without fragrance, scentless
ansatus -a -um, ansiferus -a -um having a handle
anserinus -a -um of the goose, of the meadows

ante- before-

Antennaria Feeler (projecting like a boat's yard-arm; the hairs of the pappus)

anthelminthicus -a -um vermifuge, worm-expelling

anthemi- Anthemis-, chamomile-

Anthemis Flowery (name, ανθεμις, used by Dioscorides)

-anthemus -a -um, -anthes -flowered, ανθος

Anthericum the Greek name, ανθερικος, for an asphodel

antherotes brilliant

-antherus -a -um -anthered

antho- flower- (ανθος)

Anthoceros Flower-horns (the conspicuously elongate, dark brown, bi-valved capsules)

anthora resembling *Ranunculus thora* in poisonous properties

Anthostema Floral-wreath (the heads of flowers)

Anthoxanthum Yellow-flower (the mature spikelets)

anthracinus -a -um coal-black

Anthriscus from a Greek name, αθρυσκον, for another umbellifer

anthropophagorus -a -um of the man-eaters (cannibal tomato)

anthropophorus -a -um man-bearing (flowers of the man orchid)

Anthurium Flower-tail (ουρα the tail-like spadix)

-anthus -a -um -flowered

Anthyllis Downy-flower (calyx hairs), ανθυλλις, used by Dioscorides

anti- against-, opposite-, opposite-to-, like-, false-, αντι-

Antiaris Against-association (the Javan upas tree, *Antiaris toxicaria*, reputedly causes the death of anyone who sleeps beneath it)

anticus -a -um turned inwards towards the axis

Antidesma Against-a-band

antidysentericus -a -um against dysentery (use in medical treatment)

antillarus -a -um from the Antilles, West Indies

antipolitanus -a -um from Antibes (Roman *Antipolis*)

antipyreticus -a -um against fire (the moss *Fontinalis antipyretica* was packed around chimneys to prevent thatch from igniting)

antiquorum of the ancients

antiquus -a -um ancient
Antirrhinum Nose-like
antrorsus -a -um forward or upward facing
anulatus -a -um with rings, ringed
-anus -a -um -belonging to, -having
anvegadensis -is -e see *andegavensis*
ap- without-, up-
aparine a name, απαρινη, used by Theophrastus (clinging, seizing)
apenninus -a -um of the Italian Apennines
Apera a meaningless name used by Adanson
aperti-, apertus -a -um open, bare, naked
aphaca a name, αφακη, used in Pliny for a lentil-like plant
aphan-, aphano- unseen-, inconspicuous-
aphanactis resembling the Andean genus *Aphanactis* (*Erigeron aphanactis*)
Aphananthe Inconspicuous-flower
Aphanes Inconspicuous (unnoticed), αφανης
Aphelandra Simple-male
aphthosus -a -um wine-glass-shaped
aphyllus -a -um without leaves, leafless (perhaps at flowering time)
apiatus -a -um bee-like, spotted
apicatus -a -um with a pointed tip, capped
apiculatus -a -um with a small broad point at the tip, apiculate (see Fig. 7(*e*))
apifer -era -erum bee-like, bee-bearing (flowers of the bee orchid), bee-flowered
apii- *Apium*-, parsley-
Apium a name used in Pliny for a celery-like plant. Some relate it to the Celtic 'apon', water, as its preferred habitat
apo- up-, without-, free-
Apocynum Dioscorides' name for *Apocynum venetum* (supposed to be poisonous to dogs)
apodectus -a -um acceptable
apodus -a -um without a foot, stalkless

Aponogeton Without-neighbour (see *Potamogeton*)
aporo- impenetrable-
appendiculatus -a -um with appendages, appendaged
applanatus -a -um flattened out
appressus -a -um lying close together, adpressed
appropinquatus -a -um near, approaching (resemblance to another species)
apricus -a -um sun-loving, of exposed places
aprilis of April (the flowering season)
Aptenia Wingless (the capsules)
apterus -a -um without wings, wingless
aquaticus -a -um living in water
aquatilis -is -e living under water
aquifolius -a -um with pointed leaves, spiny-leaved
Aquilegia Eagle (claw-like nectaries)
aquilinus -a -um of eagles, eagle-like (the appearance of the vasculature in the cut rhizome of *Pteridium*)
aquilus -a -um blackish-brown
arabicus -a -um, *arabus -a -um* of Arabia, Arabian
Arabidopsis *Arabis*-resembling
Arabis Arabian (derivation obscure)
arachnites spider-like, αραχνη
arachnoides, *arachnoideus -a -um* cobwebbed, covered with a weft of hairs
aragonensis -is -e from Aragon, Spain
Aralia origin uncertain, could be from French Canadian, aralie
araneosus -a -um spider-like, like a cobweb
aranifer -era -erum spider-bearing
araroba the Brazilian name for the powdery secretion produced by *Andira araroba*
araucanus -a -um from the name of a tribe in Arauco, Southern Chile
Araucaria from the Chilean name, araucanos, for the tree
Araujia from the Brazilian name for the cruel plant
arborescens becoming or tending to be of tree-like dimensions

arboreus -a -um tree-like, branched

arboricolus -a -um living on trees

arbusculus -a -um, arbuscularis -is -e small-tree-like, shrubby

arbustivus -a -um coppiced, growing with trees

Arbutus the ancient Latin name

Arceuthobium Juniper-life (European species is a parasite on *Juniperus*)

Archangelica supposedly revealed to Matthaeus Sylvaticus by the archangel as a medicinal plant

arche-, archi- primitive-, αρχη-, beginning-, original-

arct-, arcto- bear-, αρκτος, northern-,

Arctanthemum Northern-flower (*Chrysanthemum arctium*)

arcticus -a -um of the Arctic regions, Arctic

Arctium Bear (a name in Pliny, the shaggy hair)

Arctostaphylos Bear's-grapes (this is the Greek version of *uva-ursi*, giving one of the repetitive botanical binomials *Arctostaphylos uva-ursi*)

Arctotis Bear's-ear

Arctous Boreal-one, αρκτος, or That-of-the-bear (the black bearberry)

arcturus -a -um bear's tail-like

arcuatus -a -um bowed, curved, arched

ardens glowing, fiery

ardesiacus -a -um, ardosiacus -a -um slate-grey, slate-coloured

Ardisea Pointed (the acute anthers)

Aregelia for E.A. von Regel (1815–1892), of St Petersburg Botanic Garden

Aremonia derived from a Greek plant name, αργεμον, for *Agrimonia*

Arenaria Sand-dweller

arenarius -a -um, arenosus -a -um growing in sand, of sandy places

arenastrus -a -um resembling *Arenaria*

arenicolus -a -um sand-dwelling

areolatus -a -um with distinct angular spaces (in the leaves)

Argemone a name, αργεμωνη, used by Dioscorides for a poppy-like plant used medicinally as a remedy for cataract *argentatus -a -um* silvered

argenteo-, *argenteus -a -um*, *argentus -a -um* silvery

aretinus -a -um from Arezzo, Italy

argillaceus -a -um growing in clay, whitish, clay-like, of clay

argo- pure white-, silvery-

arguti- sharply saw-toothed

argutus -a -um sharply toothed or knotched, lively, piercing

Argylia for Archibald Campbell, 3rd Duke of Argyll and plant introducer

argyr-, *argyreus -a -um*, *argyro-* silvery, silver-, αργυρος

Argyranthemum Silver-flower (*Chrysanthemum*)

ari- *Arum-*

aria a name, αρια, used by Theophrastus for a whitebeam

arianus -a -um from Afghanistan, Afghan

aridus -a -um of dry habitats, dry, arid

arietinus -a -um like a ram's head, ram-horned

-aris -is -e -pertaining to

Arisaema Blood-*arum* (αιμα blood)

aristatus -a -um with a beard, awned, aristate (see Fig. 7(*g*))

Aristida Awn (the awns are conspicuous)

Aristolochia Best-childbirth, αριστολοχεια (abortifacient property)

-arius -a -um -belonging to, -having

arizelus -a -um notable

arizonicus -a -um from Arizona, USA

armatus -a -um thorny, armed

armeniacus -a -um Armenian (mistake for China), the dull orange colour of *Prunus armeniaca* fruits

armenus -a -um from Armenia, Armenian

Armeria ancient Latin name for a *Dianthus*

armillaris -is -e, *armillatus -a -um* bracelet-like

armoraceus -a -um horse-radish-like

Armoracia of uncertain meaning, a Greek name formerly used for a cruciferous plant, possibly the widespread *Raphanus raphanistrum*

armoricensis -is -e from Brittany peninsula, NW France (*Armorica*)

Arnebia from an Arabic name

Arnica Lamb's-skin (αρνακις from the leaf texture)

arnoldianus -a -um of the Arnold Arboretum, Massachusetts

Arnoseris Lamb-succour (Lamb's succory), αρνος–σερις

aromaticus -a -um fragrant, aromatic

Aronia a derivative name from *Aria*

arrectus -a -um raised up, erect

arrhen-, arrhena- male-, stamen-

Arrhenatherum Male-awn (the male lower spikelet is awned), αρρην–αθηρ

arrhizus -a -um without roots, rootless, α–ριζα (the minute, floating *Wolffia* has no roots)

Artemisia for Artemis (Diana), wife of Mausolus, of Caria, Asia Minor

arthro- joint-, jointed-, αρθρον-

Arthrocnemum Jointed-internode, αρθρον–κνημη

Arthropodium Jointed-foot (the jointed pedicels)

Arthropteris Jointed-fern (the rachis of the frond is jointed towards the base)

articulatus -a -um, arto- jointed, joint-, articulated

Artocarpus Bread-fruit (*artopta* a baker)

Arum a name, αρον, used by Theophrastus

Aruncus the Greek name

arundinaceus -a -um *Arundo*-like, reed-like

Arundinaria derived from *arundo*, a cane

Arundo the old Latin name for a reed

arvalis -is -e of arable or cultivated land

arvaticus -a -um from Arvas, N Spain

arvensis -is -e of the cultivated field, of ploughed fields

arvernensis -is -e from Auvergne, France

Asarina a Spanish vernacular name for *Antirrhinum*

Asarum a name, ασαρον, used by Dioscorides

ascendens upwards, ascending

-ascens -becoming, -turning to, -tending-towards

asclepiadeus -a -um *Asclepias*-like

Asclepias for Aesculapius, mythological god of medicine

asco- bag-like-, bag-

Asimina from the French/Italian name, asiminier

asininus -a -um ass-like (eared), loved by donkeys

Asparagus the Greek name for plants producing edible turions from the rootstock

asper -era -erum, asperi- rough (the surface texture)

asperatus -a -um rough

aspergilliformis -is -e shaped like a brush, with several fine branches

aspermus -a -um without seed, seedless

aspernatus -a -um despised

aspersus -a -um sprinkled

Asperugo Rough-one

Asperula Little-rough-one (*asper* rough)

Asphodeline *Asphodelus*-like

Asphodelus the Greek name, ασφοδελος, for *Asphodelus ramosus*

Aspidistra Small-shield (ασπιδιον), the stigmatic head

Aspidium Shield, ασπις (the shape of the indusium)

aspleni- *Asplenium*-, spleenwort-

Asplenium Without-spleen (Dioscorides' name, ασπληνον, for spleenwort)

assimilis -is -e resembling, like, similar to

assurgens, assurgenti- rising upwards, ascending

Aster Star, αστηρ

-aster -ra -rum -somewhat resembling (usually implying inferiority), -false

asterias star-like

asterioides Aster-like

asthmaticus -a -um of asthma (its medicinal use)

astictus -a -um immaculate, without blemishes, unspotted

Astilbe Without-brilliance (στιλβω), the flowers

Astragalus Ankle-bone, a Greek name in Pliny for a plant with vertebra-like knotted roots

Astrantia l'Ecluse's name for masterwort
astro- star-shaped-
Astrocarpus Star-fruit
Astrophytum Star-plant (the plant's shape)
atamasco a vernacular name
-ater, -atra, -atrum -matt-black
athamanticus -a -um, athemanticus -a -um of Mt Athamas, Sicily
Athanasia Immortal (without death, funerary use of *Tanacetum*)
athero- bristle-
Athyrium Sporty, αθυρω (from the varying structure of the sori)
-aticus -a -um, -atilis -is -e -from (a place)
atlanticus -a -um of the Atlas Mountains, North Africa, of Atlantic areas
atomerius -a -um speckled
Atraphaxis an ancient Greek name, ατραφαξυς, for *Atriplex*(q.v.)
atratus -a -um blackish, clothed in black
atri-, atro- very dark-, better- (a colour)
Atriplex the ancient Greek name, ατραφαξυς, used by Pliny
Atropa Inflexible (one of three Fates, Ατροπος)
atrovirens very dark green
attenuatus -a -um tapering, drawn out to a point
-atus -a -um -having, -rendered
Aubrieta (Aubretia) for Tournefort's artist friend, Claude Aubriet (1668–1743)
Aucuba latinized Japanese name, aokiba
aucuparius -a -um bird-catching, of bird-catchers (fruit used as bait, *avis capio*)
augescens increasing
augurius -a -um of the soothsayers
augustus -a -um stately, noble, tall
aulicus -a -um courtly, princely
aulo-, aulaco- tube-, furrowed-
aurantiacus -a -um, aurantius -a -um orange-coloured

aurarius -a -um, auratus -a -um golden, ornamented with gold
aureliensis -is -e from Orleans, France
aureo-, aureus -a -um golden-yellow
auricomus -a -um with golden hair
Auricula l'Ecluse's name, *auricula-ursi*, for the bear's-ear *Primula*
auricula-judae Jew's-ear (the shape of the fruiting body of *Auricularia auricula-judae*)
auriculatus -a -um lobed like an ear, with lobes, ear-shaped
aurigeranus -a -um from Ariège, France
auritextus -a -um cloth-of-gold
auritus -a -um with ears, long-eared
aurorius -a -um orange
aurosus -a -um golden
australasiacus -a -um, australiensis -is -e Australian, South Asiatic
australis -is -e southern, of the South
austriacus -a -um from Austria, Austrian
austro- southern
autumnalis -is -e of the autumn (flowering or growing)
avellanus -a -um from Avella, Italy or hazy
Avena Nourishment
avenaceus -a -um oat-like
Averrhoa for Ibn Rushd Averrhoes, 12th century Arabian physician, translator of Aristotle's work
Avicennia for Ibn Sina (Avicenna) (980–1037), Arabian philosopher and physician
avicularis -is -e of small birds, eaten by small birds
avium of the birds
axillaris -is -e arising from the leaf axils (flowers), axillary
Axyris Without-edge, αξυρης (the bland flavour)
Azalea Of-dry-habitats (αζαλεος, formerly used for *Loiseleuria*)
Azana the Mexican vernacular name
Azara for J.N. Azara
azarolus a vernacular name

Azolla from a South American name thought to refer to its inability to survive out of water, or αζο–ολλυμι dryness-kills

azureus -a -um sky-blue

Babiana Baboon (from the Afrikaans for baboon; they feed on the corms)

babylonicus -a -um from Babylon

baccans berried-looking (shining red to purple, berry-like fruits of *Carex baccans*)

baccatus -a -um having berries, fruits with fleshy or pulpy coats

Baccharis an ancient Greek name (doubtful etymology)

baccifer -era -erum bearing berries

bacillaris -is -e rod-like, staff-like, stick-like

badius -a -um reddish-brown

baeticus -a -um from Spain (*Baetica*), Andalusian

Baillonia for H. Baillon (1827–1895), French botanist

Balanites Acorn (the Greek name, βαλανος, describes the fruit of some species)

balansae, balansanus -a -um for Benedict Balansa, French plant collector (1825–1891)

balanus the ancient name, βαλανος, for an acorn

balcanus -a -um of the Balkans, Balkan

Baldellia for B. Bartolini-Baldelli, Italian nobleman

baldensis -is -e from the area of Mt Baldo, N Italy

baldschuanicus -a -um from Baldschuan, Bokhara

Ballota the Greek name, βαλλωτη, for *Ballota nigra*

balsamae, balsameus -a -um, balsamoides balsam-like, yielding a balsam

balsamifer -era -erum yielding a balsam, producing a fragrant resin

banaticus -a -um from Banat, Romania

Banksia, banksii, banksianus -a -um for Sir Joseph Banks (1743–1820), one time President of the Royal Society and patron of the sciences

Baphia Dye (cam-wood, *Baphia nitida*, gives a red dye, it is also used for violin bows)

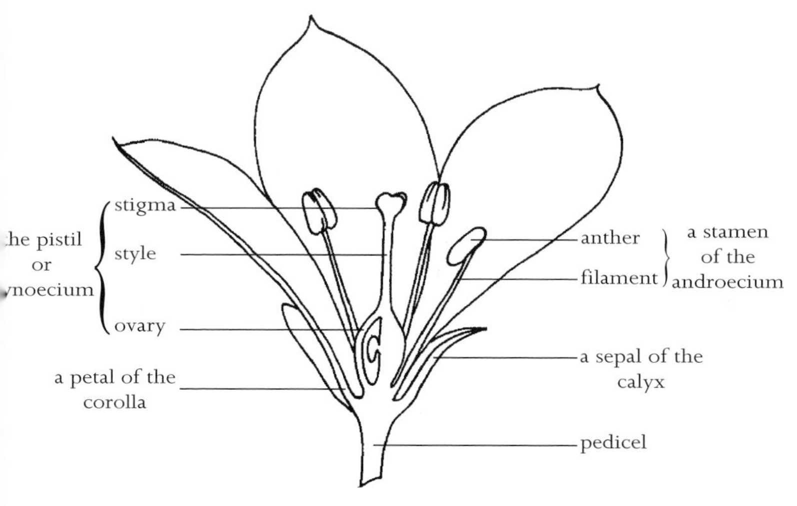

Fig. 1. The parts of a flower as seen in a stylized flower which is cut vertically in half.

baphicantus -a -um of the dyers, dyers'

Baptisia Dye (βαπτω), several yield indigo

barbadensis -is -e from Barbados, West Indies

barba-jovis Jupiter's beard

Barbarea Lyte's translation of Dodoens' *Herba Sanctae Barbarae*, for St Barbara

barbarus -a -um foreign, from Barbary (North African coast)

barbatus -a -um with tufts of hair, bearded

barbellatus -a -um having small barbs

barbi-, barbigerus -a -um bearded

barometz from a Tartar word meaning lamb (the woolly fern's rootstock)

Barosma Heavy-odour

Bartsia for Johann Bartsch, Dutch physician

bary- heavy-

basalis -is -e sessile-, basal-

Basella the Malagar vernacular name
basi-, -bassos of the base-, from the base-, βασις
basilaris -is -e relating to the base
basilicus -a -um princely, royal
bastardii for T. Bastard, author of the *Flora of Maine & Loire*, 1809
batatas Haitian name for sweet potato, *Ipomoea batatas*
bathy- thick-, deep-
batrachioides water-buttercup-like, *Batrachium*-like
Batrachium Little-frog (Greek, βατραχος, for some *Ranunculus* species)
battandieri for Jules Aime Battandier (1848–1922) of the Algiers Medical School
Bauhinia for the 16th century botanists Caspar Bauhin (1550–1624) and his brother John
baxarius -a -um clog-like
beccabunga from an old German name 'Bachbungen', mouth-smart or streamlet-blocker
Begonia for Michel Begon (1638–1710), French Governor of Canada and patron of botany
Belamcanda from an Asian name for the leopard lily
belladonna beautiful lady, the juice of the deadly nightshade was used to beautify by inducing pallid skin and dilated pupils when applied as a decoction
bellatulus -a -um somewhat beautiful
Bellevalia for P.R. Belleval (1558–1632), early plant systematist
bellidi-, bellidiformis -is -e, belloides daisy-like, *Bellis*-like
Bellis Pretty (a name used in Pliny)
bellobatus -a -um beautiful bramble
bellus -a -um handsome, beautiful, neat
benedictus -a -um well spoken of, blessed, healing
benjamina from an Indian vernacular name, ben-yan
benjan the Indian vernacular for weeping fig, *Ficus benjan*
Benthamia (Cornus) for George Bentham (1800–1884) author of *Genera Plantarum*, with Sir Joseph Hooker

Benzoin, *benzoin* from an Arabic or Semitic name, signifying perfume or gum

Berberis Barbed-berry (from medieval Latin, *barbaris*, an Arabic name for North Africa)

bergamia from the Turkish name, beg-armodi, of the bergamot orange

Bergenia for Karl August von Bergen (1704–1760), German physician and botanist

berolinensis -is -e from Berlin, Germany

Berteroa for Carlo G.L. Bertero (1789–1831), Italian physician

Bertholletia for Claude-Louis Berthollet (1748–1822), French chemist (Brazil nut)

Berula the Latin name in Marcellus Empyricus

Bessera, *besserianus -a -um* for Dr W.S.J.G. von Besser (1782–1842), Professor of Botany at Brody, Ukraine

Beta the Latin name for beet

betaceus -a -um beet-like, resembling *Beta*

betoni- *Betonica*-like-

Betonica from a name in Pliny for a medicinal plant from Vettones, Spain

betonicifolius -a -um betony-leaved

Betula Pitch (the name in Pliny, bitumen is distilled from the bark)

betulinus -a -um, *betuloides*, *betulus -a -um* *Betula*-like, birch-like

bholuo from a vernacular name for a *Daphne*, bholu swa

bi-, *bis-* two-, twice-

bialatus -a -um two-winged (usually the stem)

bicalcaratus -a -um two-spurred

bicameratus -a -um two-chambered

bicapsularis -is -e having two capsules

bicolor of two colours

Bidens Two-teeth (the scales at the fruit apex)

biennis -is -e (with a life) of two years, biennial

bifarius -a -um in two opposed ranks (leaves or flowers), two-rowed *bifidus -a -um* deeply two-cleft, bifid

bifurcatus -a -um divided into equal limbs, bifurcate

Bignonia (Bignona) for Abbé Jean Paul Bignon (1662–1743), librarian to Louis XIV of France

bijugans, bijugus -a -um two-together, yoked

bilimbi a vernacular name for the cucumber-tree (*Averrhoa bilimbi*)

-bilis -is -e -able, -capable

Billardiera, billardierei (billardierii) for Jaques Julien Houtou de la Billardière (1755–1834), French botanist

Billbergia for J.G. Billberg (1772–1844), Swedish botanist

binatus -a -um with two leaflets, bifoliate

Biophytum Life-plant (sensitive leaves)

Biscutella Two-trays (the form of the fruit)

bisectus -a -um cut into two parts

biserratus -a -um twice-toothed, double toothed (leaf margin teeth themselves toothed)

bistortus -a -um twice twisted (the roots, from the medieval name for bistort)

bisuntinus -a -um from Besançon, France

bithynicus -a -um from Bithynia, Asia Minor

bituminosus -a -um tarry, clammy, adhesive

Bixa from a South American vernacular name for *B. orellana*, the annatto tree

Blackstonia for John Blackstone, English botanical writer

blandus -a -um pleasing, alluring, not harsh, bland

-blastos, -blastus -a -um -shoot

blattarius -a -um cockroach-like, an ancient Latin name

Blechnum the Greek name, βληχνον, for a fern

blepharo- fringe-, eyelash-, βλεφαρις

blepharophyllus -a -um with fringed leaves

blitoides resembling *Blitum* (*Chenopodium*), from a plant name used by Greek and Latin writers

Blumenbachia for Johann Friedrich Blumenbach (1752–1840), medical doctor of Göttingen

Blysmus meaning uncertain, βλυζω
bocasanua -a -um from the Sierra de Bocas, Panama
Boehmeria for G.R. Boehmer (1723–1803), professor at Wittenberg
boeoticus -a -um from Boeotia, near Athens
Boerhaavia for Herman Boerhaave (1668–1739), early plant systematist
bolanderi for Prof. H.N. Bolander (1831–1897) of Geneva, plant collector in California and Oregon
Bombax Silk (the hair, kapok, covering of the seeds)
bombyci- silk- (*bombyx*, a silkworm)
bombycinus -a -um silky
bona-nox good night (night-flowering)
bonariensis -is -e from Buenos Aires, Argentina
bondus an Arabic name for a hazel-nut
bononiensis -is -e from Bologna, N Italy, or Boulogne, France
bonus-henricus good King Henry (allgood or mercury)
Boophone Ox-killer (narcotic property)
Borago Shaggy-coat (*burra* rough), the leaves
borbonicus -a -um from Reunion Island, Indian Ocean, or for the French Bourbon Kings
borealis -is -e northern, of the North
boreaui for Alexander Boreau (1803–1875), Belgian botanist
Boreava for Alexander Boreau, Belgian botanist (1803–1875)
boris-regis for King Boris
Boronia for Francesco Boroni, assistant to Humphrey Sibthorp in Greece
Borreri for W. Borrer (1781–1862), British botanist
bosniacus -a -um from Bosnia
bothrio- minutely pitted-
botry- bunched-, panicled-
Botrychium Little-bunch, βοτρυχιον, the fertile portion of the frond of moonwort
botryodes, botrys resembling a bunch of grapes
botrytis -is -e racemose, racemed, bunched

botulinus -a -um shaped like small sausages (branch segments)
Bougainvillea for Louis Antoine de Bougainville (1729–1811), French navigator
brachi-, brachy- short- (βραχυς)
brachiatus -a -um arm-like, branched at about a right-angle
brachybotrys short-clustered, shortly bunched
Brachychiton Short-tunic
Brachycome Short-hair
Brachypodium Short-foot
Brachystelma Short-crown (the coronna)
bracteatus -a -um with bracts, bracteate (as in the inflorescences of *Hydrangea*, *Poinsettia* and *Acanthus*)
bracteosus -a -um with large or conspicuous bracts
brandisianus -a -um, brandisii for Sir Dietrich Brandis (1824–1907), dendrologist of Bonn
brasiliensis -is -e from Brazil, Brazilian
Brassia for William Brass, orchidologist
Brassica Pliny's name for various cabbage-like plants
brassici- cabbage-, *Brassica-*
brevi-, brevis -is -e short-, abbreviated-
breviscapus -a -um short-stalked, with a short scape
Briza Food-grain (an ancient Greek name, βριζα, for rye)
Bromus Food (the Greek name, βρωμος, for an edible grass)
-bromus -smelling, -stinking
bronchialis -is -e throated, of the lungs (medicinal use)
Broussonetia for T.N.V. Broussonet (1761–1807), French naturalist
Browallia for Bishop John Browall
brumalis -is -e of the winter solstice, winter-flowering
Brunfelsia for Otto Brunfels (1489–1534), who pioneered critical plant illustration
brunneus -a -um russet-brown
Brunnichia for M.T. Brunnich, 18th century Scandinavian naturalist
Bryanthus Moss-flower
bryoides moss-like

Bryonia Sprouter (a name, βρυωνια, used by Dioscorides)
Bryum Moss, from the Greek βρυον
bubalinus -a -um, bubulinus -a -um of cattle, of oxen
buboni- of the groin (βουβων groin)
buccinatorius -a -um, buccinatus -a -um heralded, trumpet-shaped, horn-shaped, trumpeter
bucephalus -a -um bull-headed
bucerus -a -um ox-horn-shaped
Buchanani either for Francis Buchanan Hamilton of Calcutta Botanic Garden or for John Buchanan, specialist on New Zealand plants
bucharicus -a -um from Bokhara, Turkestan
bucinalis -is -e, bucinatus -a -um trumpet-shaped, trumpet-like
Buda an Adansonian name of no meaning
Buddleia (Buddleja) for Adam Buddle (d. 1715), English botanist
bufonius -a -um of the toad, living in damp places (*bufo* common toad)
bulbi-, bulbo- bulb-, bulbous-
bulbifer -era -erum producing bulbs (often when these take the place of normal flowers)
Bulbine the Greek name, βολβος, for a bulb
bulbocastanus -a -um chestnut-brown-bulbed
Bulbophyllym Bulb-leaved (the pseudobulbs)
bulbosus -a -um swollen, having bulbs, bulbous
bullatus -a -um with a bumpy surface, puckered, blistered, bullate
bullulatus -a -um with small bumps or blisters or bullae
bumannus -a -um having large tubercles
Bumelia an ancient Greek name for an ash tree
-bundus -a -um -having the capacity for
Bunias the Greek name, βουνιας, for a kind of turnip
Bunium a name, βουνιον, used by Dioscorides
Buphthalmum, buphthalmoides, bupthalmoides Ox-eyed (βουσ–οφθαλμος)
Bupleurum Ox-rib, an ancient Greek name, βουπλευρος, used by Nicander
burmanicus -a -um from Burma, Burmese

burnatii for Emile Burnat (1828–1920), French botanist
burs-, bursa- pouch-, purse-
bursa-pastoris shepherd's purse
bursiculatus -a -um formed like a purse, pouch-like
Butomus Ox-cutter, a name, βουτομος, used by Theophrastus with reference to the sharp-edged leaves
butyraceus -a -um oily, buttery
Butyrospermum Butter-seed (oily seed of shea-butter tree)
buxi- *Buxus*-, box-
buxifolius -a -um box-leaved
Buxus an ancient Latin name used by Virgil for *B. sempervirens*
byrs-, byrsa- pelt-, hide- (leather-)
byzantinus -a -um from Istanbul (*Byzantium* Constantinople), Turkish

Cabomba from a Guyanese vernacular name
cacao Aztec name for the cocoa tree, *Theobroma cacao*
cachemerianus -a -um, cachemiricus -a -um from Kashmir
cacti- cactus-like- (originally the Greek χαχτυς was an Old World spiny plant, not one of the *Cactaceae*)
cacumenus -a -um of the mountain top
cadmicus -a -um with a metallic appearance
caducus -a -um transient, not persisting, caducous
caeno-, caenos- fresh-, recent-
caenosus -a -um muddy, growing on mud
caerulescens turning blue, bluish
caeruleus -a -um dark sky-blue
Caesalpinia for Andreas Caesalpini (1519–1603), Italian botanist
caesi-, caesius -a -um bluish-grey, lavender-coloured
caesiomurorum of the blue walls (*Hieraceum*)
caespitosus -a -um growing in tufts, matted, tussock-forming
caffer -ra -rum, caffrorum from South Africa, of the unbelievers (Kaffirs)
cainito the West Indian name for the star apple
Caiophora Burn-carrier (the stinging hairs)

cairicus -a -um   from Cairo, Egypt
cajan, Cajanus   from the Malay name, katjan, for the pigeon pea
cajennensis -is -e   from Cayenne, French Guyana
cajuputi   the Malayan name
Cakile   from an Arabic name
cala-   beautiful-
calaba   the West Indian name
calabricus -a -um   from Calabria, Italy
Caladium   from the Indian name, kaladi, for an elephant ears *Arum*
Calamagrostis   Reed-grass, name, καλαμος–αγρωστις, used by Dioscorides
calamarius -a -um   reed-like, resembling *Calamus*
calami-   *Calamus*-, reed-
calaminaris -is -e   cadmium-red, growing on the zinc ore, calamine
Calamintha   Beautiful-mint, καλος–μινθη
calamitosus -a -um   causing loss, dangerous, miserable
Calamondin   a name for the fruit of × *Citrofortunella*
Calamus   the name, καλαμος, for a reed
Calanthe   Beautiful-flower
calanthus -a -um   beautiful-flowered
Calathea   Basket-flower (the inflorescence)
calathinus -a -um   basket-shaped, basket-like
calcaratus -a -um, calcatus -a -um   spurred, having a spur
calcareus -a -um   of lime-rich soils, chalky
calcar-galli   cock's-spur
calceolatus -a -um   shoe-shaped, slipper-shaped
calceolus -a -um   like a small shoe
calcicolus -a -um   living on limy soils
calcifugus -a -um   disliking lime, avoiding limy soils
calcitrapa   caltrop (the fruit's resemblance of the spiked ball used to damage the hooves of charging cavalry horses)
calcitrapoides   *Centaurea*-like, resembling *Calcitrapa* (for *Centaurea*)
caledonicus -a -um   from Scotland (Caledonia), Scottish, of northern Britain

calenduli- *Calendula*-, marigold-

Calendula First-day-of-the-month (Lat. *calendae* associated with paying accounts and settling debts), for its long flowering period

Calepina an Adansonian name perhaps relating to Aleppo

calidus -a -um fiery, warm

californicus -a -um from California, USA

caliginosus -a -um of misty places

Calla Beauty (a name used in Pliny)

calli-, callis- beautiful- (καλλι-)

Calliandra Beautiful-stamens (shaving-brush tree)

callibotryon beautifully bunched

Callicarpa Beautiful-fruit (its metallic-violet drupes)

callifolius -a -um *Calla*-leaved

callimorphus -a -um of beautiful form or shape

Callistemon Beautiful-stamens (bottle-brush tree)

Callistephus Beautiful-crown (the flower-heads)

callistus -a -um very beautiful

Callitriche Beautiful-hair, καλλιτριχον

callizonus -a -um beautifully zoned (colouration)

callosus -a -um hardened, with a hard skin

Calluna Sweeper (former common use as brooms)

callybotrion fine-racemed

calo- beautiful- (καλος)

Calochortus Beautiful-grass (the grass-like foliage)

Calodendrum (-on) Beautiful-tree

calomelanos beautifully-dark

Caloncoba Beautiful-*Oncoba*

calophrys with dark margins

calostomus -a -um beautiful mattress (growth habit)

calpetanus -a -um from Gibraltar

calpophilus -a -um estuary-loving, estuarine

Caltha old Latin name, used by Pliny for a marigold

calvescens with non-persistent hair, becoming bald

calvus -a -um naked, hairless, bald

calyc-, calyci- calyx-, καλυξ

Calycanthus Calyx-flower (the undifferentiated tepals of the spiral perianth of allspice)

calycinus -a -um, calycosus -a -um with a persistent calyx, calyx-like

Calycocarpum Cup-fruit (the concavity on one side of the stone)

Calycotome Split-calyx (the upper part of the calyx splits before anthesis)

calyculatus -a -um with a small calyx, resembling a small calyx

calyptr-, calyptro- hooded-, lidded-

calyptratus -a -um with a cap-like cover over the flowers or fruits

Calystegia Calyx-cover (the calyx is at first obscured by prophylls)

camaldulensis -is -e from the Camaldoli gardens near Naples

camara a West Indian name, arched

-camarus -a -um -chambered

Camassia a North American Indian name, quamash, for an edible bulb

cambodgensis -is -e from Cambridge

cambodiensis -is -e from Cambodia, SE Asia

cambrensis -is -e, cambricus -a -um from Wales (Cambria), Welsh

Camelina Dwarf-flax, χαμαι–λινον

Camellia for George Joseph Kamel, or Cameli (1661–1706), Jesuit traveller and plant illustrator

camelliiflorus -a -um *Camellia*-flowered

camelorus -a -um of camels (they feed upon the camel thorn, *Alhagi camelorum*, also known as the manna plant because of the crust of dried honey-sap forming on the leaves overnight)

cammarus -a -um lobster (from a name used by Dioscorides)

campani- bell-

Campanula Little-bell

campanularius -a -um, campanulatus -a -um, campanulus -a -um bell-shaped, bell-flowered

campanus -a -um from Campania, Italy

campester -tris -tre of the pasture, from flat land, of the plains

camphoratus -a -um camphor-like scented

Camphorosma Camphor-odour (the fragrance)
Campsis Curvature, καμπε (the bent stamens)
campto- (kampto-) bent-
Camptosorus Curved-sorus
Camptostylus Bent-style (the long curved style)
Camptothecium Bent-theca (the curved capsule)
camptotrichus -a -um with curved hairs
campyl-, campylo- bent-, curved-
Campylopus Curved-stalk
Campylotropis Curved-keel (the curved, rostrate keel petals)
camtschatcensis -is -e, camtschaticus -a -um from the Kamchatka Peninsula, E Siberia
camulodunum from Colchester
canadensis -is -e from Canada, Canadian
canaliculatus -a -um furrowed, channelled
cananga from a Malayan name
canariensis -is -e from the Canary Isles, of bird food
canarinus -a -um yellowish, resembling *Canarium*
canarius -a -um canary-yellow
cancellatus -a -um cross-banded, chequered, latticed
candelabrus -a -um candle-tree, like a branched candlestick
candicans whitish, hoary-white, with white woolly hair
candidus -a -um shining-white
Candollea for August Pyramus de Candolle (1778–1841), Professor of Botany at Geneva
canephorus -a -um like a basket bearer
canescens turning hoary-white, off-white
caninus -a -um of the dog, sharp-toothed or spined, wild or inferior, not of cultivation
Canna Reed (*canna*)
cannabinus -a -um hemp-like, resembling *Cannabis*
Cannabis Dioscorides' name, καννaβις, for hemp
cano- hairy-
cantabricus -a -um from Cantabria, N Spain

cantabrigiensis -is -e from Cambridge (Cantabrigia)
cantianus -a -um from Kent, England
Cantua from a Peruvian vernacular name
canus -a -um whitish-grey, white
capax wide, broad
capensis -is -e from Cape Colony, South Africa
caperatus -a -um wrinkled
capillaceus -a -um, capillaris -is -e, capillatus -a -um hair-like, very slender
capilliformis -is -e hair-like
capillipes with a very slender stalk
capillus-veneris Venus' hair
capitatus -a -um growing in a head, head-like (inflorescence)
capitellatus -a -um growing in a small head
capitulatus -a -um having small heads
capnoides smoke-coloured (καπνοειδες)
cappadocicus -a -um, cappadocius -a -um from Cappadocia, Asia Minor
capraeus -a -um, capri- of the goat, goat-like (smell), *capraea*a she-goat
capreolatus -a -um tendrilled, with tendrils, twining
capreolus the roe deer
caprifolium Goat-leaf (an old generic name)
Capriola Goat
Capsella Little-case (the form of the fruit)
Capsicum Biter (καπτη), the hot taste
capsularis -is -e producing capsules
caput-galli cock's-head
caput-medusae Medusa's-head
caracalla beautiful snail, cloaked
caracasanus -a -um from Caracas
Caragana the Mongolian name, caragan, for the plant
carambola a vernacular name for the carambola-tree (*Averrhoa carambola*)
carataviensis -is -e from Karatau, Kazakhstan

Cardamine Dioscorides' name, καρδαμινη, for cress
Cardaminopsis *Cardamine*-resembler
cardamomum ancient Greek name for the Indian spice
Cardamon the Greek name, καρδαμον, for garden cress
Cardaria Heart-like (the fruiting pods)
cardi-, *cardio-* heart-shaped-, καρδια
cardiacus -a -um of heart conditions (medicinal use), καρδιακος
cardinalis -is -e cardinal-red
cardiopetalus -a -um with heart-shaped petals
Cardiospermum Heart-seed; refers to the white, heart-shaped aril on the black seeds
cardui- *Carduus-*, thistle-
carduncculus -a -um thistle-like
Carduus Thistle (a name in Virgil)
Carex Cutter (the sharp leaf margins of many)
caribaeus -a -um from the Caribbean
Carica From-*Carya* (mistakenly thought to be the provenance of the pawpaw, *Carica papaya*)
carici-, *caricinus -a -um*, *caricosus -a -um* sedge-like, resembling *Carex*
caricus -a -um from Caria, province of Asia Minor
carinatus -a -um keeled, having a keel-like ridge
carinthiacus -a -um from Carinthia, Austria
Carlina for Charlemagne (Carolinus); his army was supposed to have been cured of the plague with a species of *Carlina*
Carmichaelia for Captain Douglas Carmichael (1722–1827), plant hunter
carmineus -a -um carmine
Carnegiea for the philanthropist Andrew Carnegie
carneus -a -um, *carnicolor* flesh-coloured
carniolicus -a -um from Carniola, former Yugoslavia
carnosulus -a -um somewhat fleshy
carnosus -a -um fleshy, thick and soft-textured
carolinianus -a -um, *carolinus -a -um* of North or South Carolina, USA
carota the old name for carrot (*Daucus carota*)

carpathicus -a -um, carpaticus -a -um from the Carpathian Mountains
Carpenteria for William M. Carpenter (1811–1848), Professor at Louisiana
carpetanus -a -um from the Toledo area of Spain
carpini- hornbeam-like-
Carpinus the ancient Latin name for hornbeam, some derive it from Celtic for a yoke
carpo-, carpos-, -carpus -a -um (karpo-) fruit-, -fruited, -podded (καρπος)
Carpobrotus Edible-fruit, καρπος–βρωτος
Carpodetus Bound-fruit (external appearance of the putaputawheta fruit)
Carrichtera for Bartholomaeus Carrichter, physician to Emperor Maximillian II
Carthamus Painted-one (Hebrew, qarthami, an orange-red dye is made from *Carthamnus tinctorius*)
carthusianorum of the Grande Chartreuse Monastery of Carthusian Monks, Grenoble, France
cartilagineus -a -um, cartilaginus -a -um cartilage-like (texture of some part, e.g. leaf margin)
Carum Dioscorides' name, καρω, for caraway
carunculatus -a -um with a prominent caruncle (seed coat outgrowth, usually obscuring the micropyle)
carvi (carui) from Caria, Asia Minor
Carya ancient Greek name, καρια, for a walnut and the tree
caryo- (karyo-) nut-, clove-, καρυον
Caryolopha Nut-crest (they form a ring)
caryophyllaceus -a -um, caryophylleus -a -um resembling a stitchwort, clove-pink-coloured
Caryopteris Nut-winged (the fruit-body splits into four, winged nutlets)
cashemirianus -a -um see *cachemerianus*
Casimiroa for Casimiro Gomez de Ortega (1740–1818), Spanish botanist

caspicus -a -um of the Caspian area

Cassia a name, κασια, used by Dioscorides from a Hebrew name (quetsi'oth) used by Linnaeus for *C. fistula* (medicinal senna)

Cassinia for Count A.H.G. de Cassini (1781–1832), French botanist

cassioides resembling *Cassia*

Cassiope mother of Andromeda in Greek mythology

Cassipourea from a vernacular name from Guyana

cassubicus -a -um from Cassubia, part of Pomerania

Castalia Spring-of-the-Muses, on Mt Parnassus

Castanea old Latin name for the sweet chestnut, from the Greek καστα

castaneus -a -um, castanus -a -um chestnut-brown

Castanopsis Chestnut-like

castello-paivae for Baron Castello de Paiva

castus -a -um spotless, pure

cat-, cata-, cato- below-, outwards-, downwards-, from-, under-, against-, along-

Catabrosa Eaten (the appearance of the tip of the lemmas, and also much liked by cattle), καταβρωσις, to swallow

catacosmus -a -um adorned

catafractus -a -um, cataphractus -a -um enclosed, armoured, closed in, mail-clad

Catalpa, catalpa from an East Indian vernacular name

Catananche Driving-force (καταναvγκε), its use in love potions by Greek women

Catapodium Minute-stalk (the spikelets are subsessile)

catappa from a native East Indian name for olive-bark tree

cataria of cats, old name for catmint (catnip)

catarractae, catarractarum growing near waterfalls, resembling a waterfall

catawbiensis -is -e from the Catawber River, Carolina, USA

catechu a vernacular name, caycao, for the betel (*Acacia catechu*) in Cochin China

catenarius -a -um, catenatus -a -um chain-like, linked

catharticus -a -um purgative, purging, cathartic

cathayanus -a -um, cathayensis -is -e from China (Cathay)

catholicus -a -um of Catholic lands (Spain and Portugal), world-wide, universal

Cattleya for William Cattley, English plant collector, and patron of Botany

Caucalis old Greek name, καυκαλις, for an umbelliferous plant

caucasicus -a -um from the Caucasus, Caucasian

caudatus -a -um, caudi- -tailed (see Fig. 7(*a*))

caudiculatus -a -um with a thread-like caudicle or tail

caulescens having a distinct stem, beginning to stem, καυλος

cauliatus -a -um, -caulis -is -e, -caulo, -caulos of the stem or stalk, -stemmed, -stalked

cauliflorus -a -um bearing flowers on the main stem, flowering on the old woody stem

causticus -a -um with a caustic taste (mouth-burning)

cauticolus -a -um growing on cliffs, cliff-dwelling

cautleoides resembling *Cautlea*

cavernicolus -a -um growing in caves, cave-dwelling

cavernosus -a -um full of holes

cavus -a -um hollow, cavitied

cayennensis -is -e from Cayenne, French Guyana

Ceanothus the ancient Greek name, κεανοθος

Cecropia for Cecrops, legendary King of ancient Athens

Cedrela Cedar-like (the wood is similar)

Cedrus the ancient Greek name, κεδρος, for a resinous tree with fragrant wood

Ceiba from a vernacular South American name for silk-cotton tree

celastri- *Celastrus*-like-

Celastrus Theophrastus' name, κηλαστον, for an evergreen tree, possibly an *Ilex*

celebicus -a -um from the Indonesian island of Celebes

celeratus -a -um hastened

-cellus -a -um -lesser, -somewhat

Celosia Burning (from κελος, for the burnt or dry flowers of some)
celtibiricus -a -um from central Spain
Celtis ancient Greek name, κελτις, for a tree with sweet fruit. Linnaeus applied this to the European hackberry
cembra the old name for the arolla or stone pine
cembroides, cembrus -a -um resembling *Pinus cembra*
cenisius -a -um from Mt Cénis on the French/Italian border
ceno-, cenose- empty-, fruitless-, κενος
Centaurea Centaur (mythical creature with the body of a horse replacing the hips and legs of a man, the name used by Hippocrates)
centaureoides resembling *Centaurea*
Centaurium for the Centaur, Chiron, who was fabled to have used this plant medicinally
centi- one hundred-, many-
centra-, centro-, -centrus -a -um spur-, -spurred, κεντρον
centralis -is -e in the middle, central
Centranthus (Kentranthus) Spur-flower
centratus -a -um many-spined, many-spurred
cepa, cepae- the old Latin name, *caepa*, for an onion, onion-
cepaeus -a -um grown in gardens, κηπος, from the ancient Greek for a salad plant
cephal- head-, head-like-, κεφαλη
Cephalanthera Head-anther (its position on the column)
Cephalanthus Head-flower (flowers are in axillary globose heads)
Cephalaria Head (the capitate inflorescence)
cephalidus -a -um having a head
cephalonicus -a -um from Cephalonia, one of the Ionian Islands
Cephalotaxus Headed-yew (the globose heads of staminate 'flowers')
cephalotes having a small head-like appearance
cephalotus -a -um with flowers in a large head
-cephalus -a -um -headed
cerae- waxy-

-ceras -horned, -podded
ceraseus -a -um waxy
cerasifer -era -erum bearing cherries (cherry-like fruits)
cerasinus -a -um cherry-red
Cerastium Horned (the fruiting capsule's shape)
Cerasus from an Asiatic name for the sour cherry
cerato- horn-shaped-, κερατος
Ceratochloa Horned-grass (the lemmas are horn-like)
Ceratonia Horned (the fruit shape of the carob)
Ceratophyllum Horn-leaf (the stag's-horn shape of the leaf)
Ceratopteris Horned fern
Ceratostigma Horned-stigma (the shape of the stigmatic head)
Cerberus Poisonous-one, after Cerberus, the three-headed guardian dog of Hades
Cercidiphyllum *Cercis*-leaved
Cercis the ancient Greek name, κερκις
Cercocarpus Tail-fruit (the persistent, long, plumose style on the fruit)
cerealis -is -e for Ceres, the goddess of agriculture
Cereus, cereus -a -um waxy (*cereus*, a wax taper)
cerifer -era -erum wax-bearing
cerinus -a -um waxy
cernuus -a -um drooping, curving forwards
Ceropegia Fountain-of-wax (appearance of the inflorescence)
Ceroxylon Wax-wood
cerris the ancient Latin name, *cerrus*, for turkey oak
cervianus -a -um of the hind or stag (*Mollugo cervianus*)
cervicarius -a -um constricted, keeled
cervinus -a -um tawny, stag-coloured
cespitosus -a -um growing in tufts – see *caespitosus*
Cestrum an ancient Greek name, κεστρον, of uncertain etymology
Ceterach an Arabic name, chetrak, for a fern
cevisius -a -um closely resembling
ceylanicus -a -um from Ceylon

chaeno- splitting-, gaping-, χαινω

Chaenomeles Gaping-apple

Chaenorrhinum Gaping-nose (analogy with *Antirrhinum*)

chaero- pleasing-, rejoicing-

Chaerophyllum Pleasing-leaf (the ornamental foliage)

chaeto- long hair-like-, χαιτη

chaixii for Abbé Dominique Chaix (1731–1800) a collaborator of Villars

chalcedonicus -a -um from Chalcedonia, Turkish Bosphorus

chamae- on-the-ground-, lowly-, low-growing-, prostrate-, false-, χαμαι

Chamaebatia Dwarf-bramble

Chamaecyparis Dwarf-cypress

Chamaedaphne Ground-laurel

chamaedrys ground oak

Chamaemelum Ground-apple (the habit and fragrance), chamomile

Chamaenerion Dwarf-oleander. Gesner's name for rosebay willow-herb

Chamaepericlymenum Dwarf-climbing-plant

chamaeunus -a -um lying on the ground

Chamomilla Dioscorides' name, χαμαιμηλος, for a plant smelling of apples

Characias the name in Pliny for a spurge with very caustic latex

charantius -a -um graceful

charianthus -a -um with elegant flowers

Charieis Elegant

-charis -beauty

chartaceus -a -um parchment-like

chasmanthus -a -um having open flowers

chathamicus -a -um from the Chatham Islands

chauno- gaping-, χαινω

Cheilanthes Lip-flower (the false indusium of the frond margin covers the marginal sori)

cheilanthus -a -um with lipped flowers

cheilo- lip-, lipped-
cheir- red (from Arabic)
Cheiranthus Red-flower (from an Arabic name for wallflower)
cheiri, cheiri- red-flowered, wallflower-
cheiro- hand-, hand-like-
Chelidonium Swallow-wort (Dioscorides' name, χελιδον, Greek for a swallow; flowering at the time of their migratory arrival)
Chelone Turtle-like (χελωνε) the turtle's-head-like corolla
Chenopodium Goose-foot (the shape of the leaves)
cherimola a Peruvian-Spanish name
Cherleria for J.H. Cherler, son-in-law of C.H. Bauhin
chermisinus -a -um red
chia from the Greek Island of Chios
Chiastophyllum Crosswise-leaf (the phyllotaxy)
chilensis -is -e from Chile, Chilean
chiloensis -is -e from Chiloe Island, off Chile
-chilos, -chilus -a -um -lipped
chima-, chimon- winter-
chimaera monstrous, fanciful
Chimaphila Winter-love (wintergreen)
Chimonanthus Winter-flower (χειμα, χειμων winter)
chinensis -is -e from China, Chinese, see *sinensis*
chio-, chion-, chiono- snow- (χιων)
chioneus -a -um snowy
Chiogenes Snow-offspring (the white berries)
Chionanthus Snow-flower (its abundant white flowers)
Chionodoxa Glory of the snow (very early flowering)
chiro- hand-
Chironia, chironius -a -um after Chiron, the centaur of Greek mythology who taught Jason and Achilles the medicinal use of plants
chirophyllus -a -um with hand-shaped leaves
-chiton -covering, -protection, -tunic (χιτων, a coat of mail)

chlamy-, *chlamydo-* cloak-, cloaked- (χλαμυς mantle or cloak)
Chlidanthus Luxurious-flower (χλιδε luxury)
chlor-, *chloro-*, *chlorus -a -um* yellowish-green- (χλωρο)
Chlora Greenish-yellow-one
chloracrus -a -um with green tips, green-pointed
chloranthus -a -um green-flowered
Chloris for Chloris, Greek goddess of flowers
chlorophyllus -a -um green-leaved
chocolatinus -a -um chocolate-brown
Choisya for Jacques Denis Choisy (1799–1859), Swiss botanist
choli- bile-like (χοληκος bile)
chondro- rough-, angular-, lumpy-, coarse- (χονδρος grain)
chordatus -a -um cord-like
chordo- string-, slender-elongate-
chori- separate-, apart-
Chorispora Separated-seed (winged seeds are separated within the fruit)
-chromatus -a -um, *-chromus -a -um* -coloured (χρωμα)
chrono- time-
chrys-, *chryso-* golden- (χρυσος)
Chrysalidocarpus Golden-fruit
Chrysanthemum Golden-flower (Dioscorides' name for *C. coronarium*). Now treated as several new genera such as *Ajania*, *Arctanthemum*, *Argyranthemum*, *Dendranthema*, *Leucanthemella*, *Leucanthemopsis*, *Leucanthemum*, *Nipponanthemum*, *Pyrethropsis*, *Rhodanthemum* and *Tanacetum*
chrysanthus -a -um golden-flowered, χρυσος–ανθεμον
chryseus -a -um golden-yellow (χρυσος gold)
Chrysobalanus Golden-acorn (the fruit of some is acorn-like)
Chrysocoma Golden-hair (the terminal inflorescence)
chrysographes marked with gold lines, as if written upon in gold
chrysolectus -a -um finishing up yellow, yellow at maturity
chrysomallus -a -um with golden wool, golden-woolly-hairy
chrysops with a golden eye

chrysopsidis -is -e resembling *Chrysopsis* (former North American generic name)
Chrysosplenium Golden-spleenwort (used for diseases of the spleen)
chrysostomus -a -um with a golden throat
Chrysothamnus Golden-shrub (its appearance when in full flower)
chrysotoxus -a -um golden-arched
-chthon-, chthono- -ground, earth- (χθωυ earth)
chyllus -a -um from a Himalayan vernacular name
chylo- sappy- (χυλος juice)
cibarius -a -um edible
cicatricatus -a -um marked with scars (left by falling structures such as leaves)
Cicenda an Adansonian name with no obvious meaning
cicer, cicerus -a -um the old Latin name, *cicer*, for the chick-pea
Cicerbita Italian name for *Sonchus oleraceus*, from an old Latin name for a thistle
Cichorium Theophrastus' name, κιχωριον
ciconius -a -um resembling a stork's neck
Cicuta the Latin name for *Conium maculatum*
cicutarius -a -um resembling *Cicuta*, with large two- or three-pinnate leaves
ciliaris -is -e, ciliatus -a -um, ciliosus -a -um fringed with hairs, ciliate
cilicicus -a -um from Cilicia, southern Turkey
-cillus -a -um -lesser
Cimicifuga Bug-repeller (*cimex*, a bug)
Cinchona for the Countess of Chinchon, wife of the Viceroy of Peru. She was cured of fever with the bark, source of quinine, in 1638, and introduced it to Spain in 1640
cincinnatus -a -um with crisped hairs
cinctus -a -um, -cinctus -a -um girdled, -edged
cineraceus -a -um, cinerarius -a -um, cinerescens ash-coloured, covered with ash-grey felted hairs
Cineraria Ashen-one (the foliage colour)
cinereus -a -um ash-grey

cinnabarinus -a -um cinnabar-red

cinnamomeus -a -um, *cinnamonius -a -um* cinnamon-brown, endearing (Ovid)

Cinnamomum the Greek name, κινναμομον, used by Theophrastus

cio- erect- (κιον)

Circaea for the enchantress Circe, κιρκε, of mythology (Pliny's name for a charm plant)

circinalis -is -e, *circinatus -a -um* curled round, coiled like a crozier, circinate

circum- around-

cirratus -a -um, *cirrhatus -a -um*, *cirrhiferus -a -um* having or carrying tendrils

cirrhosus -a -um tawny-coloured (κιρρος tawny)

Cirsium the ancient Greek name, κιρσιον, for a thistle

Cissus the ancient Greek name, κισσ, for ivy

cisti- *Cistus*-like-

Cistus Capsule, κισθυς (conspicuous in fruit)

citratus -a -um *Citrus*-like

citreus -a -um, *citrinus -a -um* citron-yellow

citri- citron-like-

citriodorus -a -um citron-scented, lemon-scented

Citrulus Little-orange (the fruit colour)

Citrus from the ancient Latin name, *citrus*

clad-, *clado-* shoot-, branch-, of the branch-

Cladium Small-branch

Cladothamnus Branched-shrub (the much-branched habit)

Cladrastis Fragile-branched (the brittle branches)

clandestinus -a -um concealed, hidden, secret

clandonensis -is -e from Clandon, Surrey

Claoxylon Brittle-wood

Clarkia for Captain William Clark

clarus -a -um clear

clausus -a -um shut, closed

clavatus -a -um, *clavi-*, *clavus -a -um* clubbed, club-shaped *claviculatus -a -um* having tendrils, tendrilled

clavigerus -a -um club-bearing

Claytonia for John Clayton (1686–1773), British botanist in America

cleio-, cleisto- shut-, closed-

Cleistanthus Hidden-flower (concealed by prominent, hairy bracts)

Cleistopholis Closed-scales (the arrangement of the inner petals)

Clematis the Greek name, κλεματις, for several climbing plants

clematitis -is -e vine-like, with long vine-like twiggy branches

Clematoclethra Climbing-*Clethra* (resembles *Clethra* but climbs like *Clematis*)

Clerodendron (um) Fortune-tree (early names for Ceylonese species *arbor fortunata* and *arbor infortunata*)

Clethra ancient Greek name, κληθρη, for alder (similarity of the leaves of some)

Clianthus Glory-flower (χλιδη glory)

Clino- prostrate-, bed-

Clinopodium Bed-foot (Dioscorides' name, κλινοποδιον, for the shape of the inflorescence)

clipeatus -a -um shield-shaped

Clivia for Lady Charlotte Clive, wife of Robert Clive of India

clivorum of the hills

Clutia (Cluytia) for Outgers Cluyt (*Clutius*) (1590–1650), of Leyden

clymenus -a -um from an ancient Greek name (see *periclymenum*)

clypeatus -a -um, clypeolus -a -um like a Roman shield

Clypeola (Clipeola) Shield (the shape of the fruit)

-cnemis, cnemi-, cnemido- -covering (ancient Greek, κνημις, for a greave or legging)

-cnemius -calf-of-the-leg, internodes, ancient Greek, κνημο

-cnemum -the-internode (Theophrastus used κναμα, κνημη tibia, for the part of the stem between the joints)

cneorum of garlands, the Greek name for an olive-like shrub

Cnicus the Greek name, κνηκος, of a thistle used in dyeing

co-, col-, con- together-, together with-, firmly-

coacervatus -a -um clustered, in clumps

coadunatus -a -um united, held-together

coaetaneus -a -um ageing together (leaves and flowers both senesce together)

coagulans curdling

coarctatus -a -um pressed together, bunched, contracted

coca the name used by South American Indians

cocciferus -a -um, coccigerus -a -um bearing berries

coccineus -a -um (cochineus) crimson (the dye produced from galls on *Quercus coccifera*)

Cocculus Small-berry (diminutive of κοκκος)

coccum scarlet

-coccus -a -um -berried (κοκκος)

Cochlearia Spoon (via latinization of German Löffelkraut, cochlear, shape of the basal leaves)

cochlearis -is -e spoon-shaped

cochleatus -a -um twisted like a snail-shell, cochleate

cochlio-, cochlo- spiral-, twisted-

cocoides *Cocos*-like, coconut-like

Cocos from the Portuguese, coco, for monkey, the features of the end of the fruit

Cadiaeum from a Malayan vernacular name, kodiho

-codon -bell, -mouth, κωδων

Codonopsis Bell-like (flower shape)

coelestinus -a -um, coelestis -is -e, coelestus -a -um sky-blue, heavenly

coeli- sky-blue-, heavenly-

coeli-rosa rose of heaven

coelo- hollow-, κοιλος

Coelocaryon Hollow-nut (the cavity in the seed)

Coeloglossum Hollow-tongue (the lip of the flower)

coen-, coenos- common-

coerulescens bluish

coeruleus -a -um blue

Coffea from the Arabic name

coggygria the ancient Greek name for *Cotinus*

cognatus -a -um closely related to

Coix the ancient Greek name, κωηξ, for Job's tears grass

Cola from the Mende, West African name, ngolo

Colchicum Colchis, a Black Sea port, used by Dioscorides as a name for *C. speciosum*

colchicus -a -um from Colchis, the Caucasian area once famous for poisons

coleatus -a -um sheath-like

coleo- sheath-, κολεος

Coleus Sheath (the filaments around the style)

coliandrus -a -um coriander-like

coll-, -collis -is -e -necked

Colletia for Philibert Collet (1643–1718), French botanist

collinus -a -um of the hills, growing on hills

colocynthis ancient Greek name, κολοκινθυς, for the cucurbit *Citrullus colocynthis*

colombinus -a -um dove-like

colonus -a -um forming a mound, humped

colorans, coloratus -a -um, -color coloured

colubrinus -a -um snake-like

columbarius -a -um, columbrinus -a -um dove-like, dove-coloured, of doves, pigeon's

Columella for the 1st century Roman writer on agriculture

columellaris -is -e having or forming small pillars

columnaris -is -e pillar-like, columnar

Columnea for Fabio Colonna of Naples (1567–1650), publisher of *Phytobasanos*

colurna the ancient name for Turkish hazel (*Corylus colurna*)

-colus -a -um -loving, -inhabiting, -dwelling (follows a place or habitat)

Colutea an ancient Greek name, κολουτεα, used by Theophrastus for a tree

com- with-, together with-

comans, comatus -a -um   hairy-tufted, hair-like

Comarum   from Theophrastus' name, κομαρος, for the strawberry tree (their similar fruits)

comaureus -a -um   with golden hair, golden-haired

Combretodendron   *Combretum*-like-tree

Combretum   a name used by Pliny for an undetermined climbing plant

Commelina   for Caspar Commelijn (1667–1731), Dutch botanist

commixtus -a -um   mixed together, mixed up

communis -is -e   growing in clumps, gregarious, common

commutatus -a -um   changed, altered (e.g. from previous inclusion in another species)

comorensis -is -e   from Comoro Islands, off Mozambique, East Africa

comosus -a -um   shaggy-tufted, with tufts formed from hairs or leaves or flowers, long-haired

compactus -a -um   close-growing, closely packed together, dense

compar   well-matched

complanatus -a -um   flattened out upon the ground

complexus -a -um   encircled, embraced

compositus -a -um   with flowers in a head, *Aster*-flowered, compound

compressus -a -um   flattened sideways (as in stems), pressed together

Comptonia   for Henry Compton (1632–1713), bishop of Oxford

comptus -a -um   ornamented, with a head-dress

con-   with-, together with-

concatenans, concatenatus -a -um   joined together, forming a chain

concavus -a -um   basin-shaped, concave

conchae-, conchi-   shell-, shell-like-

conchifolius -a -um   with shell-shaped leaves

concinnus -a -um   well-proportioned, neat, elegant, well-put-together

concolor   uniformly-coloured, coloured similarly

condensatus -a -um   crowded together

conduplicatus -a -um   twice-pleated, double-folded (e.g. aestivation of *Convolvulus*)

condylodes   knobbly, with knuckle-like bumps, κονδυλος

confertus -a -um crowded, pressed-together
confluans flowing-together
confluentes from Koblenz, Germany
conformis -is -e symmetrical, conforming to type or relationship
confusus -a -um easily mistaken for another species, intricate
congestus -a -um arranged very close together, crowded
conglomeratus -a -um clustered, crowded together
conicus -a -um cone-shaped, conical
conifer -era -erum cone-bearing
conii- hemlock-like, resembling *Conium*
Conium the Greek name, κωνειον, for hemlock plant and poison
conjugalis -is -e, conjugatus -a -um joined together in pairs, conjugate
conjunctus -a -um joined together
connatus -a -um united, joined
connivens converging, connivent
cono- cone-shaped-, κωνος
conoides, conoideus -a -um cone-like
Conophytum Cone-plant (its inverted conical habit)
Conopodium Cone-foot
conopseus -a -um cloudy, gnat-like
Conringia for Hermann Conring, German academic
consanguineus -a -um closely related, of the same blood
consimilis -is -e much resembling
Consolida Make-whole (the ancient Latin name from its use in healing medicines)
consolidus -a -um stable, firm
conspersus -a -um speckled, scattered
conspicuus -a -um easily seen, marked, conspicuous
constrictus -a -um erect, dense
contemptus -a -um despising, despised
contiguus -a -um close and touching, closely related
contorus -a -um twisted, bent
contra-, contro- against-
contractus -a -um drawn together

conterminus -a -um closely related, close in habit or appearance
contortus -a -um twisted
controversus -a -um doubtful, controversial
Convallaria Of-the-valley (the natural habitat of lily-of-the-valley)
convalliodorus -a -um lily-of-the-valley-scented
conversus -a -um turning towards, turning together
convexus -a -um humped, bulged outwards, convex
convolutus -a -um rolled together
Convolvulus Interwoven (a name in Pliny)
Conyza a name, κονυζα, used by Theophrastus
copallinus -a -um from a Mexican name, yielding copal-gum
cophocarpus -a -um basket-fruited
copiosus -a -um abundant, copious
Coprosma Dung-smelling (the odour of the plant)
copticus -a -um from Coptos, near Thebes, Egyptian
coracensis -is -e from Korea, Korean
coracinus -a -um raven-black
coralliferus -a -um coral-bearing
corallinus -a -um, corallioides coral-red, κοραλλιον, coral-like
Corallorhiza Coral-root (the rhizomes)
corbariensis -is -e from Corbières, France
corbularia like a small basket
Corchorus the Greek name for jute
cordatus -a -um, cordi- heart-shaped, cordate (see Fig. 6(*e*))
cordiacus -a -um cordial
cordifolius -a -um with heart-shaped leaves
Cordyline Club (κορδυλε), some have large club-shaped roots
coreanus -a -um from Korea, Korean
Corema Broom (Greek name suggested by the bushy habit)
Coreopsis Bug-like (κορις) the shape of the fruits
coriaceus -a -um tough, leathery, thick-leaved
Coriandrum Theophrastus' name, κοριανδρον, for *C. sativum*
Coriaria Leather (*corium* leather) used in tanning
coriarius -a -um of tanning, leather-like, of the tanner

corid- *Coris*-like

corii- leathery-

coritanus -a -um resembling *Coris*, from the East Midlands (home of the *Coritani* tribe of ancient Britons)

corneus -a -um horny

corni- horned-, horn-bearing-, *Cornus-*

cornicinus -a -um horny-skinned or coated

corniculatus -a -um having small horn- or spur-like appendages or structures

cornifer -era -erum, corniger -era -erum, -cornis -is -e horned, horn-bearing

cornifolius *Cornus*-leaved

cornubiensis -is -e from Cornwall (*Cornubia*), Cornish

cornucopiae horn-of-plenty, horn-full

Cornus Horn (the ancient Latin name for the cornelian cherry, *Cornus mas*)

-cornus, cornutus -a -um horn-shaped, -horned

Corokia from a New Zealand Maori vernacular name

corollinus -a -um with a conspicuous corolla

Coronaria Crown-material (used in making chaplets)

coronarius -a -um garlanding, forming a crown

coronatus -a -um crowned

Coronilla Little-crown (the arrangement of the flowers)

Coronopus Theophrastus' name, κορωνη–πους, for crowfoot (leaf-shape)

Corrigiola Shoe-thong (the slender stems)

corrugatus -a -um wrinkled, corrugated

corsicus -a -um from Corsica, Corsican

Cortaderia Cutter (from the Spanish-Argentinian name for *Cortaderia selloana*, which refers to the sharp-celled margins of the leaves)

corticalis -is -e, corticosus -a -um with a notable, pronounced or thick bark

coryandrus -a -um with helmet-shaped stamens

Corydalis Crested-lark (the spur of the flowers)

corylinus -a -um, *coryli-* hazel-like, resembling *Corylus*

Corylopsis Hazel-resembler

Corylus Helmet (the Latin name for the hazel)

corymbosus -a -um with flowers arranged in corymbs, with a flat-topped raceme (see Fig. 2(*d*))

coryne-, *coryno-* club-, club-like-

Corynephorus Club-bearer (κορυνηφορος the awns)

corynephorus -a -um clubbed, bearing a club

coryph- at the summit-

corys-, *-corythis -is -e* helmet-, -cucculate (Greek κορις)

Cosmos Beautiful (κοσμος)

-cosmus -a -um -beauty, -decoration

costalis -is -e, *costatus -a -um* with prominent ribs, with a prominent mid-rib

Cotinus ancient Greek name (κοτινος) for a wild olive

Cotoneaster Quince-like (the leaves of some species are similar to quince, *cotoneum*)

Cotula Small-cup (κοτυλη), the leaf arrangement

Cotyledon Cupped (the leaf shape)

coulteri for Thomas Coulter (1793–1843), Irish physician and botanist

coum from a Hebrew name

cous Coan, from the island of Cos, Turkey

cracca name used in Pliny, for a vetch

Crambe ancient Greek name, κραμβη, for a cabbage-like plant

crassi- thick-, fleshy-

crassicaulis -is -e thick-stemmed

Crassula Succulent-little-plant (*crassus* thick)

crassus -a -um thick, fleshy

Crataegus Strong (the name, κραταιος, used by Theophrastus), the timber

Crataeva for Crateva, an ancient Greek botanist

Crateranthus Bowl-flower (the shape of the corolla tube)

crateri-, cratero- strong-, goblet-shaped-, a cup
crateriformis -is -e goblet- or cup-shaped, with a shallow concavity
creber -ra -rum, crebri- densely clustered, frequently
crenati-, crenatus -a -um with small rounded teeth (the leaf margins, see Fig. 4(*a*))
crepidatus -a -um sandal- or slipper-shaped
Crepis a name, κρηπις, used by Theophrastus, meaning not clear
crepitans rattling (as the seeds in the pod of the sandbox tree, *Hura crepitans*), rustling
Crescentia for Pietro de Crescenzi (1230–1321), of Bologna
cretaceus -a -um of chalk, inhabiting chalky soils
creticus -a -um from Crete, Cretan
crini- hair-, *crinis-*
criniger -era -erum carrying hairs
Crinitaria Long-hair (the inflorescence)
crinitus -a -um with long soft hairs
Crinodendron Lily-tree (κρινον), floral similarity
Crinum Lily (κρινον)
crispatus -a -um closely waved, curled
crispus -a -um with a waved or curled margin
crista-galli cock's comb (the crested bracts)
cristatus -a -um tassel-like at the tips, crested
Crithmum Barley (the similarity of the seed) κριθη
croaticus -a -um from Croatia
crocatus -a -um citron-yellow, saffron-like (used in dyeing)
croceus -a -um saffron-coloured, yellow
Crocosmia Saffron-scented (the dry flowers) κροκος–οσμη
crocosmifolius -a -um with *Crocosmia*-like leaves
Crocus Saffron, from the Chaldean name
Crossandra Fringed-anther (κροσσος)
Crotalaria Rattle (seeds loose in the inflated pods of some)
Croton Tick (the seeds of some look like ticks)
Crotonogyne Female-*Croton*

Crucianella Little-cross (= *Phuopsis*)
Cruciata Cross (Dodoens' name refers to the arrangement of the leaves)
cruciatus -a -um arranged cross-wise (leaf arrangement)
crucifer -era -erum cross-bearing, cruciform
cruentatus -a -um stained with red, bloodied
cruentus -a -um blood-coloured, bloody, blood-red
crumenatus -a -um pouched
crura, cruris legged, leg, shin
crus leg, shin
crus-andrae St Andrew's cross
crus-galli cock's spur or leg
crus-maltae, crux-maltae Maltese cross
crustatus -a -um encrusted
cruzianua -a -um from Santa Cruz
crypt-, crypto- obscurely-, hidden-, κρυπτος
Cryptanthus Hidden-flower, the concealed flowers of Earth star
Cryptogramma(e) Hidden-lines (κρυπτος–γραμμη), the concealed lines of sori
Cryptomeria Hidden-parts (the inconspicuous male cones)
crystallinus -a -um with a glistening surface, as though covered with crystals
Ctenanthe Comb-flower, the bracteate flower-head
Ctenitis Little-comb κτεις, κτενος
Ctenium Comb (the one-sided, awned, spike-like inflorescence)
cteno-, ctenoides comb-like-, comb- (κτεινος)
Ctenolophon Comb-crest (the comb-like aril of the seed)
cubeba a local vernacular name
cubitalis -is -e a cubit tall (the length of the forearm plus the hand)
Cucurbita the Latin name for the bottle-gourd, *Lagenaria*
Cucubalus a name in Pliny
cuculi of the cuckoo
cucullaris -is -e, cucullatus -a -um hood-like, hooded
cucumerinus -a -um resembling cucumber, cucumber-like

cucurbitinus -a -um   melon- or marrow-like, gourd-like
cujete   a Brazilian name
culinaris   of food, of the kitchen
cultoris, *cultorus -a -um*   of gardeners, of gardens
cultratus -a -um, *cultriformis -is -e*   shaped like a knife-blade
cultus -a -um   cultivated, grown
-culus -a -um   -lesser
cumulatus -a -um   piled-up, enlarged, perfect
-cundus -a -um   -dependable, -able
cuneatus -a -um, *cuneiformis -is -e*   narrow below and wide above, wedge-shaped
Cunninghamia   for J. Cunningham, discoverer in 1702 of *C. lanceolata* in Chusan, China
Cunonia   for J.C. Cuno (1708–1780), Dutch naturalist
Cuphea   Curve (κυφος), the fruiting capsule's shape
cupreatus -a -um   coppery, bronzed
cupressinus -a -um, *cupressoides*   cypress-like, resembling *Cupressus*
Cupressus   Symmetry (the conical shape), in mythology Apollo turned Kypressos into an evergreen tree
cupreus -a -um   copper-coloured, coppery
cupularis -is -e   cup-shaped
curassavicus -a -um   from Curaçao, West Indies
curcas   ancient Latin name for *Jatropha*
Curculigo   Weevil (the beak of the fruit)
Curcuma   the Arabic name for turmeric
curti-, *curto-*, *curtus -a -um*   shortened-, short
curtisiliquus -a -um   short-podded
curvatus -a -um, *curvi-*   curved
curvidens   with curved teeth
Cuscuta   the medieval name for dodder
cuspidatus -a -um, *cuspidi-*   abruptly narrowed into a short rigid point (cusp), cuspidate
cutispongeus -a -um   spongy-barked (*Polyscias cutispongea* is the Sponge-bark tree)

cyaneus -a -um, cyano- Prussian-blue, κυανεος dark-blue
Cyanotis Blue-ear
cyanus blue (an old generic name)
Cyathea Little-cup (the basin-like indusium around the sorus)
cyathophorus -a -um cup-bearing
cybister tumbler-shaped
Cycas Theophrastus' name, κοικας, for an unknown palm
Cyclamen Circle (the twisted fruiting stalk)
cyclamineus -a -um resembling *Cyclamen*
cycl-, cyclo- circle-, circular-
cyclius -a -um round, circular
cyclops gigantic (one-eyed giants of Greek mythology)
cydoni-, cydoniae- *Cydonia-*, quince-
Cydonia the Latin name for an 'apple' tree from Cydon, Crete
cylindricus -a -um, cylindro- long and round, cylindrical
Cymbalaria Cymbal (κυμβαλον), the peltate leaves
cymbalarius -a -um cymbal-like (the leaves of toadflax)
cymbi-, cymbidi- boat-shaped-, boat- (κυμβε)
Cymbidium Boat-like (the hollow recess in the lip)
cymbiformis -is -e boat-shaped
Cymbopogon Bearded-cup
cymimum an old generic name, *cumin*
cymosus -a -um having flowers borne in a cyme (see Fig. 3(*a–d*))
cynanchicus -a -um of quinsy (literally dog-throttling, κυναγχω), from its former medicinal use
cynanchoides resembling *Cynanchum*
Cynanchum Dog-strangler (some are poisonous)
Cynapium Dog-parsley (implying inferiority)
cyno- dog-, κυων (usually has derogatory undertone, implying inferiority)
cynobatifolius -a -um eglantine-leaved
cynoctonus -a -um dog's-bane
Cynodon Dog-tooth (the form of the spikelets)
Cynoglossum Hound's-tongue

cynops the ancient Greek name, κυωνοπς, for a plantain
Cynosurus Dog-tail
cyparissias cypress-leaved (used in Pliny for a spurge)
Cyperus the Greek name, κυπειρος, for several species
Cypripedium Aphrodite's-slipper (Kypris was a name for Aphrodite or Venus)
Cyrilla for Dominica Cyrillo (1734–1799), Professor of Medicine at Naples
cyrt- curved-, arched- (κυρτος)
Cyrtogonone an anagram of *Crotonogyne*, a related genus
Cyrtomiun Bulged, κυρτωμα (the leaflets)
cyst-, *cysti-*, *cysto-* hollow-, pouched-
Cystopteris Bladder-fern (κυστις), from the inflated-looking indusia
Cytisus the Greek name, κυτισος, for a clover-like plant

Daboecia (Dabeocia) for St Dabeoc, Welsh missionary to Ireland
Dacrydium Little-tear (δακρυδιον), its exudation of small resin droplets
dactyl-, *dactylo-*, *-dactylis*, *dactyloides* finger-, δακτυλος, finger-like-
Dactylis Grape-bunch (the inflorescence)
Dactylorchis Finger orchid (the arrangement of the root-tubers)
Dahlia for Anders Dahl, who studied under Linnaeus
dahuricus -a -um, *dauricus -a -um*, *davuricus -a -um* from Dauria, NE Asia, near Chinese border
dalmaticus -a -um from Dalmatia, eastern Adriatic, Dalmatian
damascenus -a -um from Damascus, coloured like *Rosa damascena*
Damasonium a name in Pliny for *Alisma*
Danaë after the daughter of Acrisius Persius, in Greek mythology
Danaea (Danaa) for J.P.M. Dana (1734–1801), Italian botanist
danfordiae for Mrs C.G. Danford
danicus -a -um from Denmark, Danish
Danthonia for Etienne Danthoine, student of the grasses of Provence, France

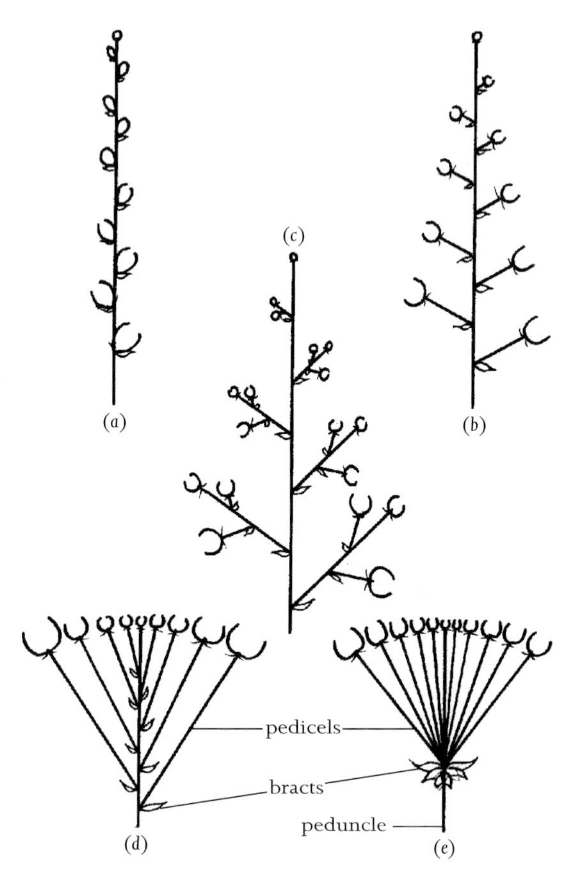

Fig. 2. Types of inflorescence which provide specific epithets.
(*a*) A spike (e.g. *Actaea spicata* L. and *Phyteuma spicatum* L.);
(*b*) a raceme (e.g. *Bromus racemosus* L. and *Sambucus racemosa* L.);
(*c*) a panicle (e.g. *Carex paniculata* L. and *Centaurea paniculata* L.);
(*d*) a corymb (e.g. *Silene corymbifera* Bertol. and *Teucrium corymbosum* R.Br.);
(*e*) an umbel (e.g. *Holosteum umbellatum* L. and *Butomus umbellatus* L.).
In these inflorescences the oldest flowers are attached towards the base and the youngest towards the apex.

danuviensis -is -e from the upper Danube

Daphne old name for bay-laurel, from that of a Dryad nymph in Greek mythology

Daphniphyllum *Daphne*-leaved

daphnoides resembling *Daphne*

Darlingtonia for C.D. Darlington, cytologist and Professor of Botany at Oxford

Darmera for Darmer (formerly *Peltiphyllum peltatum*)

Darwinia for Dr Erasmus Darwin (1731–1802), author of *The Botanic Garden* and grandfather of Charles R. Darwin (1809–1882)

darwinii for Charles Robert Darwin (1809–1882), naturalist and evolutionist, author of *The Origin of Species by means of Natural Selection*

dasy- thick-, thickly-hairy-, woolly-, δασυς

dasyclados shaggy-twigged

Dasylepis Thick-scales (the clustered scales on the stout pedicels)

dasyphyllus -a -um thick-leaved

-dasys -hairy

dasytrichus -a -um thickly haired

Datura from an Indian vernacular name

dauci- carrot-like, resembling *Daucus*

Daucus the Latin name for a carrot

Davallia for Edmond Davall (1763–1798), Swiss botanist

Davidia, *davidii*, *davidianus -a -um* for l'Abbé Armand David (1826–1900), collector of Chinese plants

de- downwards-, outwards-, from-

dealbatus -a -um with a white powdery covering, white-washed, whitened

debilis -is -e weak, feeble, frail

dec-, *deca-*, *decem-* ten-, tenfold-

Decaisnea for Joseph Decaisne (1809–1882), French botanist

decalvans balding, becoming hairless

decandrus -a -um ten-stamened

deciduus -a -um not persisting beyond one season, deciduous *decipiens* deceiving, deceptive

declinatus -a -um turned aside, curved downwards

Decodon Ten-teeth (from the horn-like processes in the calyx sinuses)

decolorans staining, discolouring

decompositus -a -um divided more than once (leaf structure), decompound

decoratus -a -um, decorus -a -um handsome, elegant, decorous

decorticans, decorticus -a -um with shedding bark

decumanus -a -um (decimanus) very large (literally, one tenth of a division of Roman soldiers)

Decumaria Ten-partite (the number of floral parts)

decumbens prostrate with tips turned up, decumbent

decurrens running down, decurrent (e.g. the bases of leaves down the stem)

decussatus -a -um at right-angles, decussate (as when the leaves are in two alternating ranks)

deficiens weakening, becoming less, deficient

deflexus -a -um bent sharply backwards

defloratus -a -um without flowers, shedding its flowers

deformis -is -e misshapen, deformed

dehiscens splitting open, gaping, dehiscent

dejectus -a -um debased, low-lying

delavayanus -a -um, delavayi for l'Abbé Jean Marie Delavay (1834–1895), French missionary and collector of plants in China

delectus -a -um choice, chosen

delicatissimus -a -um most charming, most delicate

deliciosus -a -um of pleasant flavour, delicious

Delonix Conspicuous-claw (on the petals)

delphicus -a -um from Delphi, central Greece, Delphic

delphinensis -is -e from Delphi

Delphinium Dolphin (the name, δελφινον, used by Dioscorides)

deltoides, deltoideus -a -um trianglular-shaped, deltoid

demersus -a -um underwater, submerged

demissus -a -um hanging down, low, weak, dwarf

dendr-, dendri-, dendro-, -dendron (-dendrum) tree-, tree-like-, on trees-

Dendranthema Tree-flower (woody *Chrysanthemum*)

dendricolus -a -um tree-dwelling

Dendrobium Tree-dweller (epiphytic)

dendroideus -a -um, dendromorphus -a -um tree-like

densatus -a -um, densi-, densus -a -um crowded, close, dense (habit of stem growth)

dens-canis dog's tooth

dens-leonis lion's tooth

Dentaria Toothwort (the signature of the scales upon the roots)

dentatus -a -um, dentosus -a -um having teeth, with outward-pointing teeth, dentate (see Fig. 4(*b*))

dentifer -era -erum tooth-bearing

denudatus -a -um hairy or downy but becoming naked, denuded

deodarus -a -um from the Indian state of Deodar (gift of God)

deorsus -a -um downwards, hanging

deorum of the gods

depauperatus -a -um imperfectly formed, dwarfed, of poor appearance, impoverished

dependens hanging down, pendent

depressus -a -um flattened downwards, depressed

derelictus -a -um abandoned, neglected

deremensis -is -e from Derema, Tanzania

-dermis -is -a -skin, -outer-surface

descendens downwards (flowering)

Deschampsia for the French naturalist M.H. Deschamps

Descurania (Descurainia) for Francois Descourain (1658–1740), French physician

deserti-, desertorus -a -um, desertoris -is -e of deserts

desma- bundle-

Desmanthus Bundle-flower (the appearance of the inflorescence)

Desmazeria (Demazeria) for J.B.H. Desmazières (1796–1862), French botanist

Desmodium Jointed-one (the lobed fruits)
detergens delaying
detersus -a -um wiped clean
detonsus -a -um shaved, bald
deustus -a -um burned
Deutzia for Johannes van der Deutz (1743–1788), Thunberg's patron
dextrorsus -a -um twining anticlockwise upwards as seen from outside
di-, dia-, dis- two- (δις), twice-, between-, away from-, different
dia- through-, across-
diabolicus -a -um slanderous, two-horned, devilish
diacritus -a -um distinguished, separated
diadema, diadematus -a -um band or fillet, crown, crown-like
dialy- very deeply incised-, separated-
diandrus -a -um two-stamened
Dianthus Jove's-flower (a name, διοσανθος, used by Theophrastus)
Diapensia formerly an ancient Greek name for sanicle but re-applied by Linnaeus
diaphanoides resembling *Hieracium diaphanum* (in leaf form)
diaphanus -a -um transparent (leaves)
Dicentra Twice-spurred (the two-spurred flowers)
dicha-, dicho- double-, into two-
Dichaetanthera Two-spurred-stamens (the two spurs below the anthers)
Dichapetalum Two-fold-petals (the petals are deeply bifid)
Dichondra Two-lumped (the two-lobed ovary)
Dichorisandra Two-separated-men (two of the stamens diverge from the remainder)
Dichostemma Twice-wreathed (two bracts cover the flower heads)
dichotomus -a -um repeatedly divided into two equal portions, equal-branched
dichrano- two-branched-
dichranotrichus -a -um with two-pointed hairs

dichroanthus -a -um with two-coloured flowers

dichromatus -a -um, dichromus -a -um, dichrous -a -um of two colours, two-coloured

Dicksonia for James Dickson (1738–1822), British nurseryman and botanist

dicoccus -a -um having paired nuts, two-berried

Dicranium Double-headed (the peristome teeth are bifid)

Dictamnus from Mt Dicte, Crete

dictyo-, dictyon netted-, -net (δικτυον)

dictyocarpus -a -um netted-fruit

didymo-, didymus -a -um twin-, twinned-, double-, equally-divided, in pairs, διδυμος

Didymochlaena Twin-cloak (indusia attached at centre and base but free at sides and apex)

dielsianus -a -um, dielsii for F.L.E. Diels (1874–1945), of the Berlin Botanic Garden

Dierama Funnel (διεραμα) the shape of the perianth

Diervilla for Dièreville, French surgeon and traveller in Canada during 1699–1700

difformis -is -e, diformis -is -e of unusual form or shape, irregular

diffusus -a -um loosely spreading, diffuse

Digitalis Thimble (from the German 'Fingerhut')

Digitaria Fingered (the radiating spikes)

digitatus -a -um fingered, hand-like, lobed from one point, digitate

Digraphis, digraphis -is -e Twice-inscribed, with lines of two colours

dilatatus -a -um, dilatus -a -um widened, spread out, dilated

dilectus -a -um precious, valuable

dilutus -a -um washed, pale

dimidiatus -a -um with two equal parts, dimidiate

diminutus -a -um very small

dimorpho-, dimorphus -a -um two-shaped, with two forms (of leaf or flower or fruit)

Dimorphotheca Two-kinds-of-container (the fruits vary in shape)

dinaricus -a -um from the Dinaric Alps

diodon two-toothed

dioicus -a -um of two houses, having separate male and female plants, δις–οικος

Dionaea synonymous with Venus

Dioscorea for Pedanios Dioscorides of Anazarbeus, Greek military physician

Diosma Divine-fragrance

Diospyros Divine-fruit (Jove's-fruit, edible fruit)

Diotis Two-ears (the spurs of the corolla)

Dipcadi from an oriental name for *Muscari*

Dipelta Twice-shielded (the capsules are included between persistent bracts)

diphyllus -a -um two-leaved

Diplachne Double-chaff

Diplazium Duplicate (the double indusium)

Diplopappus Double-down

Diplotaxis Two-positions (διπλοος–ταξις, the two-ranked seeds)

diplotrichus -a -um, diplothrix having two kinds of hairs

dipsaceus -a -um teasel-like, resembling *Dipsacus*

Dipsacus Dropsy, διψακος (analogy of the water-collecting leaf-bases)

diptero-, dipterus -a -um two-winged

Dipteronia Twice-winged (the two-winged carpels of the fruits)

dipyrenus -a -um two-fruited, two-stoned

Dirca an ancient Greek name from mythology

dis- two-, different

Disanthus Two-flowers (the paired flowers)

Discaria Discoid (the prominent disc)

discerptus -a -um disc-like, discoid

disci-, disco- disc-

discipes with a disc-like stalk

Discoglypremna Engraved-disc-shrub (the flowers have a deeply segmented disc)

discoides discoid

discolor of different colours, two-coloured

disermas with two-glumes

disjunctus -a -um separated, not grown together, disjunct

dispar unequal, different

dispersus -a -um scattered

dissectus -a -um (disectus) cut into many deep lobes

dissimilis -is -e unlike

dissitiflorus -a -um with flowers not in compact heads

Dissomeria Two-fold-parts (the petals are twice as many as the sepals)

distachyon, distachyus -a -um two-branched, two-spiked, with two spikes

distans widely separated, distant

distichus -a -um in two opposed ranks (leaves or flowers)

distillatorius -a -um shedding drops, of the distillers

distortus -a -um malformed, grotesque, distorted

Distylium Two-styles (the conspicuous, separate styles)

distylus -a -um two-styled

diurnus -a -um lasting for one day, day-flowering, of the day

diutinus -a -um, diuturnus -a -um long-lasting

divaricatus -a -um wide-spreading, straggling, divaricate

divensis -is -e from Chester (*Deva*)

divergens spreading out, wide-spreading, divergent

diversi-, diversus -a -um differing-, variable-, diversely-

divionensis -is -e from Dijon, France

divisus -a -um divided

divulsus -a -um torn violently apart

divus -a -um belonging to the gods

Dizygotheca Two-yoked-case (the four-lobed anthers)

Docynia an anagram of *Cydonia*

dodec-, dodeca- twelve-

Dodecatheon Twelve-gods (an ancient name)

Dodonaea for Rembert Dodoens (*Dodonaeus*) (1518–1585), physician and botanical writer

dolabratus -a -um axed, axe-shaped

dolabriformis -is -e   hatchet-shaped
dolicho-   long- (δολιχος)
Dolichos   the ancient Greek name, δολιχος, for long-podded beans
dolichostachyus -a -um   long-spiked
dolobratus -a -um   hatchet-shaped, see *dolabratus*
-dolon   -net, -snare, -trap
dolosus -a -um   deceitful
domesticus -a -um   of the household
donax   an old Greek name, δοναξ, for a reed
Doritis   Lance-like (the long lip of the corolla)
Doronicum   from an Arabic name, doronigi
Dorotheanthus   Dorothea-flower (for Dr Schwantes' mother, Dorothea)
dorsi-, *-dorsus -a -um*   on the back-, -backed, outside (outer curve of a curved structure)
dortmanna   for Herr Dortmann (*c.* 1640)
-dorus -a -um   -bag-shaped, -bag
dory-   spear- (δορυ)
Dorycnium   ancient Greek name, δορυκνιον, for a *Convolvulus* re-applied by Dioscorides
Doryopteris   Spear-fern
Douglasia, *douglasii*   for David Douglas (1798–1834), plant collector in the American North-west for the RHS
-doxa   -glory
Draba   a name, δραβη, used by Dioscorides for *Lepidium draba*
drabae-, *drabi-*   Draba-like
Dracaena   Female-dragon (δρακαινα)
dracontius -a -um   dragon-like
Dracunculus   Little-dragon (a name used by Pliny)
drepanus -a -um, *drepano-*   sickle-shaped
Drepanocarpus   Curved-fruit (Leopard's claw)
Drepanocladus   Curved-branch (the arched lateral branches)
drepanus -a -um   from a town in Western Sicily
Drimia   Acrid (the pungent juice from the roots)

drimyphilus -a -um salt-loving, halophytic
Drimys Acrid (δριμυς), the taste of the bark
Drosanthemum Dewy-flower (glistens with epidermal hairs)
Drosera Dew (the glistening glandular hairs)
drucei for George Claridge Druce (1859–1932), British botanist
drupaceus -a -um stone-fruited with a fleshy or leathery pericarp, drupe-like
Dryas Oak-nymph, δρυας (the leaf shape), one of the mythological tree nymphs or Dryads
drymo- wood-, woody-
dryophyllus -a -um oak-leaved
Dryopteris Oak-nymph-fern (δρυοπτερις), Dioscorides' name for a woodland fern
Drypetes Stone-fruits (the hard seeds)
dubius -a -um uncertain, doubtful
Duchesnea for Antoine Nicolas Duchesne (1747–1827), French botanist
dulcamara bitter-sweet
dulcis -is -e sweet-tasted, mild
dumalis -is -e, dumosus -a -um compact, thorny, bushy
dumetorum of bushy habitats, of thickets
dumnoniensis -is -e from Devon, Devonian
dumosus -a -um thorn-bushy, scrubby
dunensis -is -e of sand-dunes
duplex, duplicatus -a -um growing in pairs, double, duplicate
duplicatus -a -um double, folded, twinned
duploserratus -a -um twice-serrate, with toothed teeth
duracinus -a -um hard-fruited, hard-berried, harsh-tasting
Durio from the Malaysian name for the fruit
durior, durius harder
duriusculus -a -um rather hard or rough
durmitoreus -a -um from the Durmitor Mountains, former Yugoslavia
durus -a -um hard, hardy
Durvillaea for J.S.C.D. d'Urville (1790–1842), French naval officer

Dyckia for Prince Salms Dyck (1773–1861), German writer on succulents

dys- poor-, ill-, bad-, difficult-

Dyschoriste Poorly-divided (the stigma)

dysentericus -a -um of dysentery (medicinal treatment for)

dyso- evil-smelling-

Dysodea Evil-scented

e-, ef-, ex- without-, not-, from out of- (privative)

ebenaceus -a -um ebony-like

ebenus -a -um ebony-black

eboracensis -is -e from York (*Eboracum*)

eborinus -a -um ivory-like, ivory-white

ebracteatus -a -um without bracts

Ebulus a name in Pliny for danewort

eburneus -a -um ivory-white with yellow tinge

ecae for Mrs E.C. Aitchison

ecalcaratus -a -um without a spur, spurless

Ecballium Expeller (the sensitive fruit of the squirting cucumber throws out, εκβαλλω, its seeds when touched)

ecbolius -a -um shooting out, cathartic

eccremo- pendent (εκκρεμες)

Eccremocarpus Hanging-fruit (εκκρεμος hanging)

Echeveria for Athanasio Echeverria y Godoy, one of the illustrators of *Flora Mexicana*

echinatus -a -um, echino- covered with prickles, hedgehog-like, εχινος

Echinocactus Hedgehog-cactus

Echinocereus Hedgehog-*Cereus*

Echinochloa Hedgehog-grass (the awns)

Echinodorus Hedgehog-bag (the fruiting heads of some species)

Echinopanax Hedgehog-*Panax*

Echinops Hedgehog-resembler

echioides resembling *Echium*
Echium Viper (a name, εχιον, used by Dioscorides)
eclectus -a -um picked out, selected
ecostatus -a -um without ribs, smooth (comparative state)
ect-, ecto- on the outside-, outwards-
ectophloeos living on the bark of another plant
edentatus -a -um, *edentulus -a -um* without teeth, toothless
edinensis -is -e of Edinburgh, Scotland
editorum of the editors
edo, edoensis from Tokyo (formerly Edo)
Edraianthus Sessile-flower (εδραιος sitting)
edulis -is -e of food, edible
effusus -a -um spread out, very loose-spreading, unrestrained
Eglanteria, eglanterius -a -um from a French name (eglantois or eglanties)
Ehretia for G.D. Ehret (1710–1770), botanical artist
Eichhornia (Eichornia) for J.A.F. Eichhorn (1779–1856), of Prussia
elae-, elaeo- olive- (ελαια)
Elaeagnus Olive-chaste-tree
Elaeis Oil (copious in the fruit of the oil-palm, *Elaeis guineensis*)
Elaeophorbia Olive-*Euphorbia* (the olive-like fruits)
elaphinus -a -um tawny, fulvous (ελαφι, a fawn)
elapho- stag's-
Elaphoglossum Stag's-tongue (shape and texture of the fronds)
elasticus -a -um yielding an elastic substance, elastic
elaterium Greek name, ελατηριον, for the squirting cucumber, driving away (squirting out seeds)
Elatine Little-fir-trees (a name, ελατινη, used by Dioscorides)
elatior, elatius taller
elatus -a -um exalted, tall, high
electus -a -um select
elegans, elegantulus -a -um graceful, elegant
eleo- marsh (ελωδης) cf. *heleo-*

Eleocharis (Heleocharis) Marsh-favour

Eleogiton (Heleogiton) Marsh-neighbour (in analogy with *Potamogeton*)

elephantidens elephant's tooth

elephantipes like an elephant's foot (appearance of the stem or tuber)

elephantus -a -um of the elephants

elephas elephantine

Eleusine from Eleusis, Greece

eleuther-, *eleuthero-* free- (ελευθερος)

eleutherantherus -a -um with stamens not united but free

Elisma a variant of *Alisma*

Elliottia for Stephen Elliott (1771–1830), American botanist, author of *Flora of South Carolina*

elliottii for either G.M. Scott-Elliott, botanist in Sierra Leone and Madagascar, or Capt. Elliott, plant grower of Farnborough Park, Hants

ellipsoidalis -is -e ellipsoidal (a solid of oval profile)

ellipticus -a -um about twice as long as broad, oblong with rounded ends, elliptic

-ellus -ella -ellum -lesser (diminutive ending), -ish

Elodea Marsh (growing in water)

elodes as *helodes*, of bogs and marshes (ελωδης)

elongatus -a -um lengthened out, elongated

Elsholtzia for Johann Sigismund Elsholtz (1623–1688), Prussian botanical writer

Elymus Hippocrates' name, ελυμος, for a millet-like grass

elytri- covering-

em-, *en-* in-, into-, within-, for-, not-

emarcidus -a -um limp, flaccid, withered

emarginatus -a -um notched at the apex (see Fig. 7(*h*))

emasculus -a -um without functional stamens

emblica an old generic name

Embothrium In-little-pits (εν–βοθριον) position of its anthers

emeritensis -is -e from Merida

emersus -a -um rising out (of the water)
emerus from an early Italian name for a vetch
emeticus -a -um causing vomiting, emetic
eminens noteworthy, outstanding, prominent
Emmenopteris Enduring (εμμενης)
emodensis -is -e, emodi from the western Himalayas, 'Mt Emodus', N India
Empetrum On-rocks (Dioscorides' name, εμπετρος, refers to the habitat)
enantio- opposite-, εναντιος-
Enantia Opposite (the one-seeded carpels contrasted to the usual state)
Enarthocarpus Jointed-fruit, εναρθρος–καρπος
enatus -a -um sprung-up (an organ from another, e.g. the coronna of *Narcissus*)
Encephalartos In-a-head-bread (farinaceous centre of the stem yields sago, as in sago-palms)
encephalo- in a head-
enculatus -a -um hooded
end-, endo- internal-, inside-, within-
endivia ancient Latin name for chicory (see *Intybus*)
Endodesmia Inside-bundle (the cup-like arrangement of the united stamens)
Endymion Selen's (Diana's) lover, of Greek mythology
enervis -is -e, enervius -a -um destitute of veins, apparently lacking nerves (veins)
Englera, Englerastrum, Englerella for Heinrich Gustav Adolf Engler (1844–1930), director of Berlin, Dahlem Botanic Garden
enki- -swollen-
Enkianthus Pregnant-flower (ενκυοσ), the coloured involucre full of flowers
ennea- nine-
enneagonus -a -um nine-angled
ensatus -a -um, ensi-, ensiformis -is -e sword-shaped (leaves)

-ensis -is -e -belonging to, -from, -of (after the name of a place)
entero- intestine-
ento-, endo- on the inside-, inwards-, within-
entomo- insect-
entomophilus -a -um of insects, insect-loving
ep-, epi- upon-, on-, over-, somewhat-
Epacris Upon-the-summit (some live on hilltops)
Ephedra from an ancient Greek name, εφεδρη, used in Pliny for *Hippuris* (morphological similarity)
ephemerus -a -um transient, ephemeral
ephesius -a -um from Ephesus, Turkey, site of the temple to Diana
epi- upon-, on-, επι
Epidendron (um) Tree-dweller (the epiphytic habit)
Epigaea, epigaeus -a -um Ground-lover (γαια)
epigeios, epigejos of dry earth, from dry habitats, επιγειος
epihydrus -a -um of the water surface
epilinus -a -um parasitic on flax, on *Linum*
Epilobium Gesner's name indicating the positioning of the corolla on top of the ovary
Epimedium the name, επιμηδιον, used by Dioscorides
Epipactis a name, επιπακτις, used by Theophrastus
epiphyllus -a -um upon the leaf (flowers or buds)
epiphyticus -a -um growing upon another plant
Epipogium (Epipogon) Over-beard (the lip of the ghost-orchid is uppermost)
Epipremnum On-trees (πρεμνον a tree stump)
epipsilus -a -um somewhat naked (the sparse foliage of *Begonia epipsila*)
epipterus -a -um on a wing (fruits)
epiroticus -a -um from the Epirus district of NW Greece
Episcia Shaded (επισκιος)
epistomius -a -um snouted (flowers)
epiteius -a -um annual
epithymoides thyme-like

epithymum upon thyme (parasitic)
equestris -is -e of horses or horsemen, equestrian
equi-, *equalis -is -e* equal-
equinoctialis -is -e of the equinox, opening at a particular hour of the day
equinus of the horse
Equisetum Horse-hair (a name in Pliny for a horsetail)
equitans astride, as on horseback (leaf bases of some monocots, e.g. *Iris*)
Eragrostis Love-grass (ερος)
Eranthemum Beautiful-flower
Eranthis Spring-flower (early flowering season)
erba-rotta red-herb (*Achillea*)
erectus -a -um upright, erect
erem- desert- (ερημια)
eremo- solitary- (ερημος)
eremophilus -a -um desert-loving, living in desert conditions
Eremurus Solitary-tail (the long raceme)
eri-, *erio-* woolly- (εριον)
Erica Pliny's version of an ancient Greek name, ερεικη, used by Theophrastus
ericetorum of heathland
ericinus -a -um, *ericoides*, *erici-* heath-like, resembling *Erica*
erigens rising-up (for horizontal branches which turn up at the end)
erigenus -a -um Irish-born
Erigeron Early-old-man (Theophrastus' name, εριο-γερον)
Erinacea, *erinaceus -a -um* Prickly-one
Erinus Dioscorides' name, εριvος, for an early flowering basil-like plant
erio-, *eryo-* woolly-, εριον-
Eriobotrya Woolly-cluster (heads of small flowers almost hidden by the indumentum)
Eriocaulon Woolly-stem
Eriogonum Wool-joints (the hairy jointed stems)

Eriophorum   Wool-bearer (cotton grass)
eriophorus -a -um   bearing wool
Erisma   Support
Erismadelphus   Brother-of-*Erisma* (related to *Erisma*)
eristhales   very luxuriant, *Eristhalis*-like
erithri-, erithro-   red- (see *erythro-*)
ermineus -a -um   ermine-coloured, white broken with yellow
Erodium   Heron, ερωδιος (the shape of the fruit)
Erophila   Spring-lover, εαρ–φιλος
erosus -a -um   jagged, as if nibbled irregularly, erose
erraticus -a -um   differing from the type, of no fixed habitat
-errimus -a -um   -est, -very, -the most (superlative)
erromenus -a -um   vigorous, strong, robust
erubescens   blushing, turning red
Eruca   Belch (the ancient Latin name for colewort)
Erucastrum   *Eruca*-flowered
Ervum   the Latin name for a vetch, called *Orobus* by Theophrastus
Eryngium   Theophrastus' name, ηρυγγιον, for a spiny-leaved plant
Erysimum   a name, ερυσιμον, used by Theophrastus
Erythea   for the daughter of night and the dragon Lado of mythology, one of the Hesperides
erythraeus -a -um, erythro-   red, ερυθρος
Erythronium   Red (flower colours)
Erythroxylon (um)   Red-wood
Erythrina   Red (the flowers of some species)
Escallonia   for the Spanish South American traveller named Escallon
Eschscholzia (Eschscholtzia)   for Johann Friedrich Eschscholtz (1793–1831), traveller and naturalist
-escens   -becoming, -ish, -becoming more
esculentus -a -um   fit to eat, edible by humans, esculent
estriatus -a -um   without stripes
esula   an old generic name from Rufinus
etesiae   annual (applied to herbaceous growth from perennial rootstock)

-etorus -a -um -community (indicating the habitat)
etruscus -a -um from Tuscany (Etruria), Italy
ettae for Miss Etta Stainbank
eu- well-, ευ-, good-, proper-, completely-, well-marked
Euadenia Well-marked-glands (the five lobes at the base of the gynophore)
euboeus -a -um, euboicus -a -um from the Greek island of Euboea
Eucalyptus Well-covered (ευ–καλυπτος) the operculum of the calyx conceals the floral parts at first
Eucharis Full-of-grace
euchlorus -a -um of beautiful green, true green
euchromus -a -um well-coloured
Euclidium Well-closed (the fruit)
Eucomis Beautiful-head
Eucommia Good-gum (some yield gutta-percha)
Eucryphia Well-covered (κρυφαιος) the leaves are clustered at the branch ends
eudorus -a -um sweetly perfumed
eudoxus -a -um of good character
Eugenia for Prince Eugene of Savoy (1663–1736), patron of botany
eugenioides *Eugenia*-like
Eulophia Beautiful-crest (the crests of the lip)
eulophus -a -um beautifully crested
eunuchus -a -um castrated (as in double flowers without stamens)
Euodia, euodes Fragrance
Euonymus (Evonymus) Famed (Theophrastus' name, ευωνυμος)
Eupatorium for Mithridates Eupator, King of Pontus, reputedly immune to poisons through repeated experimentation with them upon himself
euphlebius -a -um well-veined
Euphorbia for Euphorbus, who used the latex for medicinal purposes
Euphrasia Good-cheer (signature of eyebright flowers as being of use in eye lotions)

euphues well-grown
eupodus -a -um long-stalked
euprepes, eupristus -a -um comely, good-looking, ευπρεπης
Euptelea Handsome-elm
Eupteris Proper-*Pteris*
eur-, euro-, eury- wide-, broad-
europaeus -a -um from Europe, European
Eurotia Mouldy-one (the pubescence)
Euryale for one of the Gorgons of mythology (had burning thorns in place of hair)
Euryops Good-looking
-eus -ea -eum -resembling, -belonging to, -noted for
Euscaphis Good-vessel (the colour and shape of the dehiscent leathery pods)
eustachyus -a -um, eustachyon having long trusses of flowers
Euterpe Attractive (the name of one of the Muses)
evanescens quickly disappearing, evanescent
evectus -a -um lifted up, springing out
evertus -a -um overturned, expelled, turned out
Evodia (Euodia) Well-perfumed
ex- without-, outside-, over and above-
Exaculum *Exacum*-like
Exacum a name in Pliny (may be derived from an earlier Gallic word, or refer to its expulsive property)
exalbescens out of *albescens* (related to)
exaltatus -a -um, exaltus -a -um lofty, very tall
exaratus -a -um with embossed grooves, engraved
exasperatus -a -um rough, roughened (surface texture)
excavatus -a -um hollowed out, excavated
excellens distinguished, excellent
excelsior, excelsus -a -um higher, taller, very tall
excisus -a -um cut away, cut out
excoriatus -a -um with peeling bark
excorticatus -a -um without bark, stripped (without cortex)

excurrens with a vein extended into a marginal tooth (as on some leaves)
exiguus -a -um very small, meagre, poor, petty
exili-, *exilis -is -e* meagre, small, few, slender, thin
eximius -a -um excellent in size or beauty, choice, distinguished
exitiosus -a -um fatal, deadly, pernicious, destructive
Exochorda Outside-cord (the vascular anatomy of the wall of the ovary)
exoletus -a -um fully grown, mature
exoniensis -is -e from Exeter, Devon
exotericus -a -um common, external, εξωτερικος
exoticus -a -um foreign, not native, exotic, εξωτικος
expansus -a -um spread out, expanded
expatriatus -a -um without a country
explodens exploding
exscapus -a -um without a stem
exsculptus -a -um with deep cavities, dug out
exsectus -a -um cut out
exsertus -a -um projecting, protruding, held out
exsurgens lifting itself upwards
extensus -a -um wide, extended
extra -outside-, beyond-, over and above-
extrorsus -a -um directed outwards from the central axis (outwards facing stamens), extrorse
exudans producing a (sticky) secretion, exuding

faba the old Latin name for the broad bean
fabaceus -a -um, *fabae-*, *fabarius -a -um* bean-like, resembling *Faba*
Fabiana for Archbishop Francisco Fabian y Fuero
facetus -a -um elegant, fine, humorous
faenum hay, fodder
fagi-, *fagineus -a -um* beech-like, *Fagus-*
Fagopyrum(on) Beech-wheat (buckwheat is from the Dutch Boekweit)

Fagus the Latin name for the beech tree
Falcaria Sickle (the shape of the leaf-segments)
falcatus -a -um, falciarius -a -um, falcatorius -a -um, falci- sickle-shaped, falcate
falciformis -is -e sickle-like
fallax deceitful, deceptive, false
Fallugia for Virgilio Fallugi, 17th century Italian botanical writer
Faradaya for Michael Faraday (1794–1867), scientist
farcatus -a -um solid, not hollow
farfara an old generic name for butterbur
fargesii for Paul Guillaume Farges (1844–1912), plant collector in Szechwan, China
farinaceus -a -um of mealy texture, yielding farina (starch), farinaceous
farinosus -a -um with a mealy surface, mealy, powdery
farleyensis -is -e from Farley Hill Gardens, Barbados, West Indies
farnesianus -a -um from the Farnese Palace gardens of Rome
farreri for Reginald J. Farrer (1880–1920), English author and plant hunter
fasciarus -a -um elongate and with parallel edges, band-shaped
fasciatus -a -um bound together, bundled, fasciated as in the inflorescence of cockscomb (*Celosia argentea* '*cristata*')
fascicularis -is -e, fasciculatus -a -um clustered in bundles, fascicled
fastigiatus -a -um with branches erect like the main stem, fastigiate
fastuosus -a -um proud
Fatsia from a Japanese name
fatuus -a -um not good, insipid, simple, foolish
Faucaria Gullet (*fauces* throat)
faucilalis -is -e wide-mouthed
favigerus -a -um bearing honey-glands
favosus -a -um cavitied, honey-combed
febrifugus -a -um fever-dispelling (medicinal property)
fecundus -a -um fruitful, fecund
fejeensis -is -e from the Fiji Islands

Felicia for a German official named Felix, some interpret it as Cheerful

Felix fruitful

felleus -a -um as bitter as gall

felosmus -a -um foul-smelling

femina feminine

Fendlera for August Fendler (1813–1883), New Mexico naturalist and explorer

fenestralis -is -e, fenestratus -a -um with window-like holes or openings (*Ouvirandra fenestralis*)

fennicus -a -um from Finland (*Fennica*), Finnish

-fer -fera -ferum, -ferus -a -um -bearing, -carrying

ferax fruitful

fero-, ferus -a -um wild, feral

ferox very prickly, ferocious

ferreus -a -um durable, iron-hard

ferrugineus -a -um rusty-brown in colour

ferruginosus -a -um conspicuously brown, rust-coloured

fertilis -is -e heavy-seeding, fruitful, fertile

Ferula Rod (the classical Latin name)

ferulaceus -a -um fennel-like, resembling *Ferula*, hollow-

festalis -is -e, festinus -a -um, festivus -a -um agreeable, bright, pleasant, cheerful, festive

Festuca Straw (a name used in Pliny)

festus -a -um sacred, used for festivals

fetidus -a -um bad smelling, stinking, foetid

fibrillosus -a -um, fibrosus -a -um with copious fibres, fibrous

ficaria small-fig, an old generic name for the lesser celandine (the root tubers)

fici-, ficoides fig-like, resembling *Ficus*

ficto-, fictus -a -um false

Ficus the ancient Latin name for the fig

-fid, -fidus -a -um -cleft,-divided

Filago Thread (the medieval name refers to the woolly indumentum) *filamentosus -a -um, filarius -a -um, fili-* thread-like, with filaments or threads

fili- thread-like-

filicaulis -is -e having very slender stems

filicinus -a -um, filici-, filicoides fern-like

filiculoides like a small fern

filiferus -a -um bearing threads or filaments

filiformis -is -e thread-like

Filipendula Thread-suspended (slender attachment of the tubers)

Filix Latin for fern

filix-femina (foemina) female fern

filix-mas male fern

fimbriatus -a -um, fimbri- fringed

finitimus -a -um neighbouring, adjoining, related (linking related taxa)

firmus -a -um strong, firm, lasting

fissi-, fissilis -is -e, fissuratus -a -um, fissus -a -um cleft, divided

Fissidens Split-teeth (the 16 divided peristome teeth)

fissus -a -um, -fissus cleft almost to the base

fistulosus -a -um hollow, pipe-like, tubular, fistular

Fittonia for E. and S.M. Fitton, botanical writers

flabellatus -a -um fan-like, fan-shaped, flabellate

flabellifer -era -erum fan-bearing (with flabellate leaves)

flabelliformis -is -e pleated fanwise

flaccidus -a -um limp, weak, feeble, soft, flaccid

flaccus -a -um drooping, pendulous, flabby

Flacourtia for Etienne de Flacourt (1607–1661), French East India Company

flagellaris -is -e, flagellatus -a -um, flagelli- with long thin shoots, whip-like, stoloniferous

flagelliferus -a -um bearing whips (elongate stems of New Zealand trip-me-up sedge)

flagelliformis -is -e long and slender, whip-like, flagelliform

flammeus -a -um flame-red, fiery-red

flammula an old generic name for lesser spearwort

flammulus -a -um flame-coloured

flav-, flavi-, flaveolus -a -um, flavo- yellowish

flavens being yellow

flavescens pale-yellow, turning yellow

flavidus -a -um yellowish

flavus -a -um bright almost pure yellow

flexi-, flexilis -is -e pliant, flexible

flexicaulis -is -e with bending stems

flexuosus -a -um zig-zag, winding, much bent, tortuous

-flexus -a -um -turned

flocciger -era -erum, floccosus -a -um bearing a woolly indumentum which falls away in tufts, floccose

flocculosus -a -um woolly

flora the Roman goddess of flowering plants

flore-albo white-flowered

florentinus -a -um from Florence, Florentine

flore-pleno double-flowered, full-flowered (flos plenus)

floribundus -a -um abounding in flowers, freely-flowering

floridanus -a -um from Florida, USA

floridus -a -um free-flowering, flowery

florindae for Mrs Florinda N. Thompson

florulentus -a -um flowery

-florus -a -um -flowered

flos-cuculi cuckoo-flowered, flowering in the season of cuckoo song

flos-jovis Jove's flower

fluctuans inconstant, fluctuating

fluitans floating on water

fluminensis -is -e growing in running water, of the river

fluvialis -is -e, fluviatilis -is -e growing in rivers and streams, of the river

foecundus -a -um fruitful, fecund

foemina feminine

foeni- fennel-like-

Foeniculum the Latin name for fennel
foenisicii of mown hay
foenum-graecum greek-hay (the Romans used *Trigonella foenum-graecum* as fodder)
foetidus -a -um, foetens stinking, bad smelling, foetid
foliaceus -a -um leaf-like
foliatus -a -um, foliosus -a -um leafy
-foliatus -a -um -leaflets, -leafleted
folio- leaflet-
foliosus -a -um leafy, well-leaved
-folius -a -um -leaved
follicularis -is -e bearing follicles (seed capsules as in hellebores)
Fontanesia for Réné Louiche Desfontaines (1752–1833), French botanist
fontanus -a -um, fontinalis -is -e of fountains, springs or fast-running streams
Fontinalis Spring-dweller (*fontanus* a spring)
Forestiera for Charles Le-Forestier (*c.* 1800), French naturalist
forficatus -a -um scissor-shaped, shear-shaped (leaves)
formicarius -a -um relating to ants (*formica*)
-formis -is -e -resembling, -shaped, -sort, -kind
formosanus -a -um from Taiwan (Formosa)
formosus -a -um handsome, beautiful, well-formed
fornicatus -a -um arched
forrestii for George Forrest (1873–1932), plant hunter in China
forsteri, forsterianus -a -um for J.R. Forster or his son J.G.A. Forster, of Halle, Germany
Forsythia for William Forsyth (1737–1804), of Kensington Royal Gardens
fortis -is -e strong
fortunatus -a -um rich, favourite
Fortunearia, fortunei for Robert Fortune (1812–1880), Scottish plant collector for the RHS in China

Fothergillia for John Fothergill (1712–1780), English physician and plant introducer

foulaensis -is -e from Foula, Scotland

foveolatus -a -um with small depressions or pits all over the surface, foveolate

fragari-, fragi- strawberry-

Fragaria Fragrance (the fruit)

fragifer -era -erum strawberry-bearing

fragilis -is -e brittle, fragile

fragrans sweet-scented, odorous, fragrant

frainetto from a local name for an oak in the Balkans

franciscanus -a -um, fransiscanus -a -um from San Francisco, California, USA

Frangula Fragile (medieval name refers to the brittle twigs of alder buckthorn)

frangulus -a -um breakable, fragile

Frankenia for John Frankenius (1590–1661), Swedish botanist

Franklinia for Benjamin Franklin (1706–1790), inventor of the lightning conductor and American President

fraternus -a -um closely related, brotherly

fraxini-, fraxineus -a -um ash-like, resembling ash

Fraxinus ancient Latin name for the ash, used by Virgil

Freesia for Friedrich Heinrich Theodor Freese, pupil of Ecklon

Fremontodendron (Fremontia) for Maj. Gen. John Charles Fremont (1813–1890), who explored western North America

frene- strap-

fresnoensis -is -e from Fresno County, California

frigidus -a -um cold, of cold habitats, of cold regions

friscus -a -um, frisius -a -um from Friesland, Friesian

Fritillaria Dice-box (the shape of the flowers)

frondosus -a -um leafy

fructifer -era -erum fruit-bearing, fruitful

fructu- fruit-

frumentaceus -a -um grain-producing

frutescens, fruticans, fruticosus -a -um shrubby, becoming shrubby

frutecorus -a -um, fruticorus -a -um of thickets

frutex shrub, bush

fruticulosus -a -um dwarf-shrubby

fucatus -a -um painted, dyed

Fuchsia for Leonard Fuchs (1501–1566), German Renaissance botanist

fucifer -era -erum drone-bearing (*fucus* drone)

fuciflorus -a -um bee-flowered (superficial resemblance of the flower)

fuciformis -is -e, fucoides bladder-wrack-like, resembling *Fucus* (seaweed)

fugax fleeting, rapidly withering, fugacious

-fugus -a -um -banishing, -putting-to-flight, -bane

Fuirena for G. Fuiren, Danish physician

fulgens, fulgidus -a -um shining, glistening (often with red flowers)

fuliginosus -a -um dirty-brown to blackish, sooty

fullonum of cloth-fullers

fulvescens becoming tawny

fulvi-, fulvo-, fulvus -a -um tawny, reddish-yellow, fulvous

Fumaria Smoke (Dioscorides' name, καπνος, referred to the effect of the juice on the eyes being the same as that of smoke)

fumeus -a -um smoke-coloured, smoky

fumidus -a -um smoke-coloured, dull grey coloured

fumosus -a -um smoky

funalis -is -e twisted together, rope-like

funebris -is -e mournful, doleful, of graveyards, funereal

fungosus -a -um spongy, fungus-like, pertaining to fungi

funiculatus -a -um like a thin cord

funiferus -a -um rope-bearing

furcans, furcatus -a -um forked, furcate

Furcraea for A.T. Fourcroy (1755–1809), French chemist

furfuraceus -a -um scurfy, mealy, scaly

furiens exciting to madness

fuscatus -a -um   somewhat dusky-brown
fusci-, fusco-, fuscus -a -um   bright-brown, swarthy, dark-coloured
fusiformis -is -e   spindle-shaped
futilis -is -e   useless

gaditanus -a -um   from Cadiz, Spain
Gagea   for Sir Thomas Gage (1781–1820), English botanist
Gaillardia   for Gaillard de Charentonneau (Marentonneau), patron of Botany
galactinus -a -um   milky
galanthi-   *galanthus-*, snowdrop-
Galanthus   Milk-white flower (the colour), γαλα–ανθος
Galax   Milky (flower colour)
galbanifluus -a -um   with a yellowish exudate (some *Ferula* species yield gum, galbanum)
galbinus -a -um   greenish-yellow
gale   from an old English vernacular name for bog-myrtle or sweet gale
galeatus -a -um, galericulatus -a -um   helmet-shaped, like a skull-cap
Galega   Milk-promoter
galegi-   resembling *Galega*
Galeobdolon   Weasel-smell (a name, γαλεη–βδολος, used in Pliny)
Galeopsis   Weasel-like (an ancient Greek name)
gali-, galioides   *Galium*-like-
Galinsoga   for Don M. Martinez de Galinsoga, Spanish botanist
galioides   bedstraw-like, resembling *Galium*
Galium   Milk, γαλα- (the flowers of *G. verum* were used to curdle milk in cheesemaking)
gallicus -a -um   from France, French, of the cock or rooster
Galphimia   anagram of *Malpighia*
galpinii   for Ernst E. Galpin
Galtonia   for Sir Francis Galton (1822–1911), pioneer in eugenics, fingerprinting and weather charting

gambogius -a -um rich-yellow, gamboge (the resin obtained from *Garcinia gambogia*)
gamo- fused-, joined-, united-, married-, γαμος-
Gamochaeta Fused-bristles (the united pappus hairs)
Gamolepis United-scales (the involucral bracts)
-gamus -a -um -marriage
gandavensis -is -e from Ghent, Belgium
gangeticus -a -um from the Ganges region
Garcinia for Laurence Garcin, a French 18th century botanist
Gardenia for Dr A. Garden (1730–1791), American correspondent with Linnaeus
gardneri, gardnerianus -a -um for Hon. E. Gardner (Nepal) or G. Gardner (Brazil)
garganicus -a -um from Monte Gargano, S Italy
Garrya for Nicholas Garry, secretary of the Hudson's Bay Company
gaster-, gastro- belly-, bellied-, γαστηρ
Gasteria Belly (the swollen base on the corolla)
Gastridium Little-paunch (the bulging of the glumes)
Gaudinia for J.F.G.P. Gaudin (1766–1833), Swiss botanist
Gaultheria for Dr Gaulthier (1750), Canadian botanist of Quebec
Gaura Superb
gayanus -a -um for Jacques E. Gay (1786–1864), French botanist
Gaylussacia for J.L. Gay-Lussac (1778–1850), French chemist
Gazania for the Greek scholar Theodore of Gaza (1398–1478), who transcribed Theophrastus' works into Latin. Some interpret it as Riches, *gaza -ae*
geito-, geitono- neighbour-, γειτων
gelidus -a -um of icy regions, growing in icy places
Gelsemium from the Italian, gelsomine, for true jasmine
gemellus -a -um in pairs, paired, twinned
geminatus -a -um, gemini- united in pairs, twinned
gemmatus -a -um jewelled
gemmiferus -a -um, gemmiparus -a -um bearing gemmae or deciduous buds or propagules (*Brassica gemmifera* Brussels sprout)

genavensis -is -e, *genevensis -is -e*   from Geneva, Switzerland
generalis -is -e   normal, prevailing, usual
geniculi-, *geniculatus -a -um*   with a knee-like bend
Genista   a name in Virgil (*planta genista* from which the Plantagenets took their name)
genisti-   broom-like, resembling *Genista*
Gentiana   a name in Pliny (for King Gentius of Illyria, 180–167 BC)
Gentianella   *Gentian*-like
gentilis -is -e   foreign, of the same race, noble
genuinus -a -um   natural, true
geo-   on or under the earth-
geocarpus -a -um   with fruits which ripen underground
geoides   *Geum*-like
geometrizans   equal, symmetrical
geophilus -a -um   spreading horizontally, ground-loving
georgei   for George Forrest (1873–1932), collector in China
georgianus -a -um   from Georgia, USA
georgicus -a -um   from Georgia, Caucasus
Geranium   Crane (Dioscorides' name, γερανιον, refers to the shape of the fruit resembling the head of a crane)
Gerardia   for John Gerard (1545–1612), author of the *Herbal* of 1597
Gerbera   for Traugott Gerber, German traveller
germanicus -a -um   from Germany, German
germinans   sprouting
Gesneria   for Conrad Gesner (1516–1565), German botanist
-geton   -neighbour, γειτων
Geum   a classical name in Pliny
gibb-, *gibbi-*, *gibbatus -a -um*   swollen on one side, gibbous
gibberosus -a -um   humped, hunchbacked
gibbosus -a -um   somewhat swollen or enlarged on one side
gibraltaricus -a -um   from Gibraltar
giganteus -a -um   unusually large or tall, gigantic, γιγαντειος
giganthes   giant-flowered

gigas giant, γιγας

gileadensis -is -e from Gilead, an Egyptian mountain range

Gilia from a Hottentot name for a plant used to make a beverage or for Felipe Salvadore Gil (*c.* 1790), Spanish writer on exotic plants

giluus -a -um, gilvo-, gilvus -a -um dull pale yellow

gingidium from an old name, γιγγιδιον, used by Dioscorides

Ginkgo derived from a Japanese name, gin-kyo

ginnala a native name for *Acer ginnala*

giraffae of giraffes

githago from an old generic name in Pliny (green with red-purple stripes)

glabellus -a -um somewhat smooth, smoothish

glaber -ra -rum, glabro smooth, without hairs, glabrous

glaberrimus -a -um very smooth, smoothest

glabratus -a -um, glabrescens becoming smooth or glabrous

glabriusculus -a -um rather glabrous, a little glabrous

glabrus -a -um smooth, hairless

glacialis -is -e of the ice, of frozen habitats

gladiatus -a -um sword-like

Gladiolus Small-sword (the leaves)

glandulifer -era -erum gland-bearing

glandulosus -a -um full of glands, glandular

glasti- *Isatis*-, woad-like-, *glastum*

glauci-, glauco-, glaucus -a -um with a white or greyish bloom, glaucous (Latin *glaucuma*, a cataract)

glaucescens, -glaucus -a -um with a fine whitish bloom, bluish-green, sea-green, glaucous

Glaucidium *Glaucium*-like

glaucifolius -a -um with grey-green leaves

glauciifolius -a -um with leaves resembling those of horned poppy, *Glaucium*

glaucinus -a -um a little clouded or bloomed (milky)

Glaucium Grey-green, γλαυκιον (the colour of *G. corniculatum* juice)

glaucophyllus -a -um glaucous-leaved

glaucopsis -is -e glaucous-looking
Glaux a name, γλαυξ, used by Dioscorides
Glechoma Dioscorides' name, γληχων, for penny-royal
Gleditsia (Gleditschia) for Johann Gottlieb Gleditsch (*Gleditsius*) (d. 1786), of the Berlin Botanic Garden
Gleichenia for F.W. Gleichen (1717–1783), German director of Berlin Botanic Garden
Gliricidia Mouse-killer (the poisonous seed and bark)
glischrus -a -um sticky, gluey, glandular-bristly
globatus -a -um arranged or collected into a ball
globosus -a -um, globularis -is -e with small spherical parts, spherical (e.g. flowers)
Globularia Globe (the globose heads of flowers)
globulifer -era -erum carrying small balls (the sporocarps of pillwort)
globulosus -a -um small round-headed
glochi-, -glochin point-, -pointed (γλωχις)
glochidiatus -a -um burred, with short barbed detachable bristles
glomeratus -a -um collected into heads, aggregated, glomerate
glomerulans, glomerulatus -a -um with small clusters or heads
glomeri- clustered-, crowded-
gloriosus -a -um superb, full of glory
glosso-, -glottis tongue-shaped, tongued (γλοσσα)
Glossocalyx Tongue-calyx (the elongated calyx lobe)
Glossopetalon Tongue-petalled (the narrow petals)
-glossus -a -um -tongue
glumaceus -a -um with chaffy bracts, conspicuously glumed
-glumis -is -e -glumed
glutinosus -a -um sticky, viscous, glutinous
Glyceria Sweet, γλυκερος (the sweet grain of *Glyceria fluitans*)
Glycine Sweet (the roots of some species)
glyco-, glycy- sweet-tasting or -smelling
Glycyrrhiza (Glycorrhiza) Sweet-root (the rhizomes are the source of liquorice)
Glyphia Engraved (the elongate grooves on the fruit wall)

glypto- cut-into-, carved-

glyptostroboides resembling-*Glyptostrobus*

Glyptostrobus Carved-cone (appearance of female cones)

Gnaphalium Soft-down (from a Greek name for a plant with felted leaves)

Gnidia, *gnidium* the Greek name for *Daphne*, from Gnidus, Crete

Godetia for C.H. Godet (1797–1879), Swiss botanist

Goldbachia for C.L. Goldbach (1793–1824), writer on Russian medicinal plants

gompho- nail-, bolt- or club-shaped

Gompholobium Club-pod (the shape of the fruit)

Gomphrena the ancient Latin name

gongylodes roundish, knob-like, swollen, turnip-shaped, γογγυλος–ωδης

gonio-, *gono-*, *-gonus -a -um* angled-, prominently angled-

Goodyera for John Goodyer (1592–1664), English botanist

Gordonia for James Gordon (1728–1791), English nurseryman

gorgoneus -a -um gorgon-like, resembling one of the snake-haired Gorgons of mythology, γοργος, terrible

gossypi-, *gossypinus -a -um* cotton-plant-like, resembling *Gossypium*

Gossypium Soft (from an Arabic name, goz, for a soft substance)

gothicus -a -um from Gothland, Sweden

gracilescens slenderish, somewhat slender

gracili-, *gracilis -is -e* slender, graceful

gracilior more graceful

gracillimus -a -um very slender, most graceful

Graderia an anagram of Gerardia, for John Gerard

graecizans becoming widespread

graecus -a -um Grecian, Greek

gramineus -a -um grassy, grass-like

gramini-, *graminis -is -e* of grasses, grass-like

grammatus -a -um marked with raised lines or stripes (γραμματα letters)

Grammitis Short-line (sori appear to join up like lines of writing at maturity)

grammo-, *-grammus -a -um* lined-, -lettered, -outline (γραμμα)

granadensis -is -e, *granatensis -is -e* either from Granada in Spain, or from Colombia, South America, formerly New Granada

granatus -a -um pale-scarlet, the colour of pomegranate, *Punica granatum*, flowers

grandi-, *grandis -is -e* large, powerful, full-grown, showy, big

grandidens with large teeth

graniticus -a -um of granitic rocks, grained

granulatus -a -um, *granulosus -a -um* as though covered with granules, tubercled, granulate

graph-, *graphys-* marked with lines-, as though written on, γραφω

grapto- lined-

grat-, *gratus -a -um* pleasing, graceful

gratianopolitanus -a -um from Grenoble, France

Gratiola Agreeableness (medicinal effect)

gratissimus -a -um most pleasing or agreeable

graveolens strong-smelling, rank-smelling, heavily scented

Grayia for Asa Gray (1810–1888), American botanist

gregarius -a -um growing together

Grevillea for Charles F. Greville FRS (1749–1809), founder member of the RHS

Grewia for Nehemiah Grew (1641–1712), British plant anatomist

Grindelia for D.H. Grindel (1776–1836), Latvian botanist

grisebachianus -a -um, *grisebachii* for Heinrich Rudolf August Grisebach (1814–1879), Botany Professor at Göttingen

Griselinia for Francesco Griselini (1717–1783), Italian botanist

griseus -a -um (grizeus) bluish- or pearl-grey

Groenlandia for the Parisian, Johannes Groenland

groenlandicus -a -um from Greenland

grosse-, *grossi-*, *grossus -a -um* very large, thick, coarse

Grossularia from the French name, groseille, gooseberry

grossularioides, grossuloides gooseberry-like, resembling *Grossularia*
grossus -a -um large
gruinus -a -um crane-like
grumosus -a -um broken into grains, tubercled, granular
guadalupensis -is -e from Guadalupe Island off lower California, USA
Guaiacum from the South American name for the wood of life tree
guajava South American Spanish name for the guava, *Psidium guajava*
guianensis -is -e from Guiana, northern South America
guineensis -is -e from West Africa (Guinea Coast)
Guizotia for Fr P.G. Guizot (1787–1874), historian
gummifer -era -erum producing gum
gummosus -a -um gummy
Gunnera for Johann E. Gunnerus (1718–1773), Norwegian botanist and cleric
gutta drop (*Dichopsis gutta* yields a latex, gutta percha (chaoutchouc))
guttatus -a -um spotted, covered with small glandular dots
Gymnadenia Naked-gland (exposed viscidia of pollen)
gymnanthus -a -um naked-flowered
gymno- naked- (γυμνο)
Gymnocarpium Naked-fruit (sori lack indusia in oak fern)
Gymnocladus Bare-branch (foliage mainly towards the ends of the branches)
Gymnogramma Naked-line (γυμνο), the sori lack a covering indusium
Gymnomitrium Naked-turban (the peristome)
Gymnopteris Naked-fern (the linear sori do not have an indusium)
Gymnostomium the mouth of the capsule of the beardless moss lacks a fringe of teeth
gyno-, -gynus -a -um relating to the ovary, female-, -pistillate, -carpelled (γυνμ)
Gynura Female-tail (the stigma)
Gypsophila Lover-of-chalk (the natural habitat) γυψος–φιλος
gyrans revolving, moving in circles

gyro-, *-gyrus -a -um* bent-, twisted-, -round (γυρος)
gyroflexus -a -um turned in a circle
gyrosus -a -um bent backwards and forwards (cucurbit anthers)

Habenaria Thong (etymology uncertain)
habr-, *habro-* soft-, graceful-, delicate-
Habranthus Graceful-flower
hadriaticus -a -um from the shores of the Adriatic Sea
haema-, *haemalus -a -um*, *haematodes* blood-red, the colour of blood (αιμα)
haemo-, *haemorrhoidalis -is -e* blood-coloured
Haemanthus Blood-flower (the fireball lilies)
haemanthus -a -um with blood-red flowers
Haematoxylon Blood-wood (the heartwood which is the source of the red dyestuff)
Hakea for Baron Christian Ludwig von Hake (1745–1818), German horticulturalist
hal-, *halo-* saline-, salt- (αλς)
halepensis -is -e, *halepicus -a -um* from Aleppo, northern Syria
Halesia for the Rev. Dr Stephen Hales (1677–1761), writer on plants
halicacabum from an ancient Greek name, from *Halicarnassus*, Bodrum, Turkey
halimi-, *halimus -a -um* orache-like, with silver-grey rounded leaves
Halimiocistus hybrids between *Halimium* and *Cistus*
Halimione Daughter-of-the-sea, αλιμος–ωνη
Halimium has leaves resembling those of *Atriplex halimus*
Halimodendron Maritime-tree (the habitat)
haliphloeos, *haliphleos* with salt-covered bark
halo-, *halophilus -a -um* salt-loving (the habitat)
Halorrhagis Seaside-grapeseed, αλσ–ραγος
Haloxylon Salt-wood (the habitat)
hama- together with-
Hamamelis Greek name, αμαμελις, for a tree with pear-shaped fruits, possibly a medlar

hamatus -a -um, hamosus -a -um   hooked at the tip, hooked
hamatocanthus -a -um   with hooked spines
Hammarbya   for Linnaeus, who had a house at Hammarby in Sweden
hamulatus -a -um   having a small hook, clawed, taloned
hamulosus -a -um   covered with little hooks
haplo-   simple-, single-, απλος
Haplopappus   Single-down (its one-whorled pappus)
harmalus -a -um   adapting, responsive, sensitive
Harpagophytum   Grapple-plant (the fruit is covered with barbed spines)
harpe-, harpeodes   sickle- (αρπη)
harpophyllus -a -um   with sickle-shaped leaves
Harungana   from the vernacular name of the monotypic genus in Madagascar
hastati-, hastatus -a -um   formed like an arrow-head, spear-shaped (see Fig. 6(*a*)), hastate
hastifer   bearing a spear
Hebe   Greek goddess of youth, daughter of Jupiter and wife of Hercules
hebe-   pubescent-, sluggish-
hebecarpus -a -um   pubescent-fruited
hebecaulis -is -e   slothful-stemmed (prostrate stems of *Rubus hebecaulis*)
hebegynus -a -um   with a blunt or soft-styled ovary, with part of the ovary glandular-hairy (*Aconitum hebegynum*)
hebetatus -a -um   dull, blunt or soft-pointed
hebriacus -a -um   Hebrew
hecisto-   viper-like-
Hedera   the Latin name for ivy
hederaceus -a -um, hederi-   ivy-like, resembling *Hedera* (usually in the leaf-shape)
Hedychium (on)   Sweet-snow (fragrant white flowers)
Hedyotis   Sweet-ear
Hedypnois   Sweet-sleep

hedys- sweet-, of pleasant taste or smell (ηδυς)
Hedysarum an ancient Greek name, ηδυσαρον, used by Dioscorides
helena from Helenendorf, Transcaucasia
Helenium for Helen of Troy (a name, ελενιον, used by the Greeks for another plant)
heleo- marsh- (ελωδης)
Heleocharis Marsh-favour (*Eleocharis*)
heli-, helio- sun-loving-, sun- (ηλιος)
Helianthemum Sun-flower
Helianthus Sun-flower
Helichrysum Golden-sun (χρυσος)
helici- coiled like a snail-shell, twisted, ελικτος
Heliconia for Mt Helicon, Greece, sacred to the Muses of mythology
Helicteres Twisted-band (the screw-shaped carpels)
Helictotrichon (um) Twisted-hair (the awns)
Helinus Tendrilled (climbing by spiral tendrils)
Heliophila Sun-lover
helioscopius -a -um sun-observing, sun-watching (the flowers track the sun's course)
Heliotropium Turn-with-the-sun
Helipterum Sun-wing (the fruit's plumed pappus)
helix ancient Greek name, ελιξ, for twining plants
Helleborus Poison-food (the ancient Greek, ελλεβορος, name for the medicinal *H. orientalis*)
hellenicus -a -um from Greece, Grecian, Greek, Hellenic
Helminthia (Helmintia) Worm (the elongate wrinkled fruits)
helo-, helodes of bogs and marshes, ελωδης
helodoxus -a -um marsh-beauty, glory of the marsh
Helosciadium Marsh-umbel
helveticus -a -um from Switzerland, Swiss
helvolus -a -um pale yellowish-brown
helvus -a -um dimly yellow, honey-coloured, dun-coloured
Helwingia for G.A. Helwing (1666–1748), German botanical writer
Helxine a name, ελξινη, used by Dioscorides formerly for pellitory

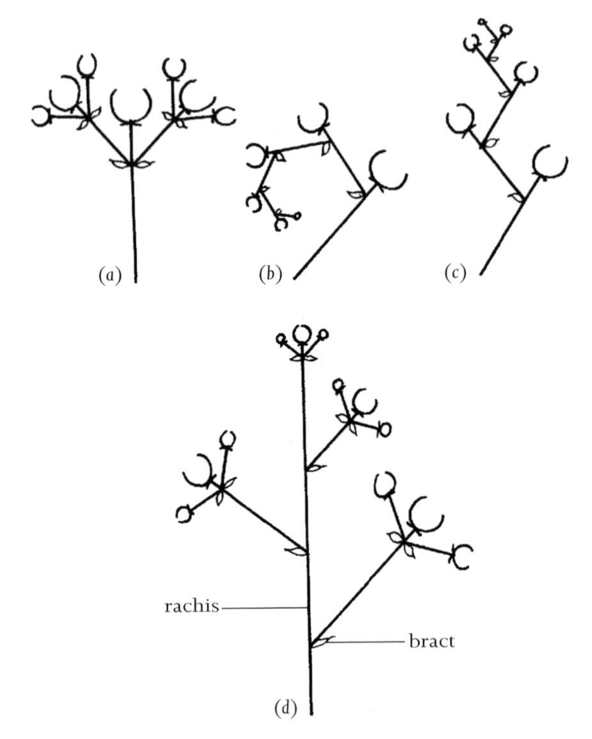

Fig. 3. Types of inflorescence which provide specific epithets.
(*a*), (*b*) and (*c*) are cymes, with the oldest flower in the centre or at the apex of the inflorescence (e.g. *Saxifraga cymosa* Waldts. & Kit.);
(*b*) may have the three-dimensional form of a screw, or bostryx;
(*c*) may be coiled, or scorpiod (e.g. *Myosotis scorpioides* L.);
(*d*) is a raceme of cymes, or a thyrse (e.g. *Ceanothus thyrsiflorus*).

Hemerocallis Day-beauty (the flowers are short-lived) (ημερα, day)
hemi- half- (ημι)
hemidartus -a -um patchily covered with hair, half-flayed
hemionitideus -a -um barren, like a mule
Hemionitis Mule (non-flowering – fern)
Hemiptelea Half-elm (πτελεα is ancient Greek for the elm)

Hemitelia Half-perfect (indusium scale-like at lower side of the sorus and caducous)
Hemizonia Half-embraced (the achenes)
henryi for Augustine Henry (1857–1930), Irish botanist
Hepatica Of-the-liver (signature of leaf or thallus shape as of use for liver complaints)
hepta-, hepto- seven-, επτα
Heracleum Hercules'-healer (a name, ηρακλειον, used by Theophrastus)
herba-barona fool's-herb (of the dunce or common man)
herba-venti wind-herb (of the steppes)
herbaceus -a -um not woody, low-growing, herbaceous
Herbertia for Dr William Herbert (1778–1847), botanist and Dean of Manchester
herco- fenced, a barrier
hercoglossus -a -um with a coiled tongue
hercynicus -a -um from the Harz Mountains, mid-Germany
hermaeus -a -um from Mt Hermes, Greece
Herminium Buttress (the pillar-like tubers), ερμις
Hermodactylus Hermes'-fingers, Ερμηϛ–δακτυλοϛ
Herniaria Rupture-wort (*hernia*), former medicinal use
herpeticus -a -um ringworm-like
hesperides of the far West (Spain)
Hesperis Evening (Theophrastus name, εσπεροϛ), also the name for Venus the evening star – that becomes Lucifer, the morning star
hespero-, hesperius -a -um western-, evening-
heter-, hetero- varying-, differing-, diversely-, other-, ετεροϛ -
Heteranthera Differing-anthers (has one large and two small)
heteronemus -a -um diverse-stemmed
heterophyllus -a -um diversely-leaved
Heteropogon Varying-beard (the twisting awns)
Heuchera for Johann Heinrich Heucher (1677–1747), German professor of medicine
Hevea from the Brazilian name, heve, for the Para-rubber tree

hex-, hexa-, hexae- six-, εξ-
hexagonus -a -um six-angled
Hexalobus Six-lobed (the six equal petals)
hexandrus -a -um six-stamened
hians gaping
hibernalis -is -e of winter (flowering or leafing)
hibernicus -a -um from Ireland (*Hibernia*), Irish
hibernus -a -um flowering or green in winter, Irish
Hibiscus an old Greek name, ηιβισκος, for mallow
hiemalis -is -e of winter
hieraci- *Hieracium-*, hawkweed-like-
Hieracium Hawkweed (Dioscorides' name, ιεραξ, for the supposed use of by hawks to give them acute sight)
Hierochloe Holy-grass, ιερος–χλοη
hierochunticus -a -um from the classical name for Jericho (*Anastatica hierochuntica* is the rose of Jericho)
hieroglyphicus -a -um marked as if with signs
hierosolymitanus -a -um from Jerusalem (the Roman name)
highdownensis -is -e connected with Sir Frederick Stern's garden at Highdown, Worthing
Hildegardia for the 11th century St Hildegard
Himantoglossum Strap-tongue (the narrow lip), ιμαντινος leather
Hippeastrum Knight-star (ηιππευς a horseman), the equitant leaves
hippo- horse-, ιππος
Hippocastanum Horse-chestnut (καστανον chestnut)
Hippocrepis Horse-shoe (κρεπις), the shape of the fruit
hippomanicus -a -um eagerly eaten by horses
hippomarathrum horse-fennel, Dioscorides' name for an Arcadian plant which caused madness in horses
Hippophae Horse-killer (used by Theophrastus for a prickly spurge)
Hippuris Horse-tail
hircinus -a -um of goats, smelling of a male goat
hirculus from a plant name in Pliny (a small goat)
Hirschfeldia for C.C.L. Hirschfeld, Austrian botanist

hirsutissimus -a -um very hairy, hairiest

hirsutulus -a -um, hirtellus -a -um, hirtulus -a -um somewhat hairy

hirsutus -a -um rough-haired, hairy

hirti-, hirtus -a -um hairy, shaggy-hairy

hirundinaceus -a -um, hirundinarius -a -um pertaining to swallows

hispalensis -is -e from Seville, southern Spain

hispanicus -a -um from Spain, Spanish, Hispanic

hispi-, hispidulus -a -um, hispidus -a -um bristly, with stiff hairs

Histiopteris Web-fern (the frond of bat-wing fern)

histrio- of varied colouring, theatrical

histrionicus -a -um of actors, of the stage

histrix showy, theatrical

Hoheria from a Maori name, houhere

Holcus Millet (the name in Pliny for a grain)

hollandicus -a -um from either northern New Guinea or Holland

Holmskioldia for Theodore Holmskjold (1732–1794), Danish botanist

holo- completely-, entirely-, entire- (ολος)

Holodiscus Entire-disc (refers to floral structure)

Holoschoenus a name, ολοσχοινος, used by Theophrastus

holosericus -a -um completely wrapped in silk

Holosteum, holostea Whole-bone (Dioscorides' name, ολος–οστεον for a chickweed-like plant)

homal-, homalo- smooth-, flat-, equal- (ομαλος)

Homalanthus Like-a-flower (the colouration of older leaves)

Homalium Equal (the petals are equal in number to the sepals – see *Dissomeria*)

Homalocephala Flat-head (the tops of the flowers)

Homaria I-meet-together (the fused filaments)

homo-, homoio-, homolo- similar-, ομοιος, not varying-, agreeing with-, uniformly-, one and the same-

Homogyne Uniform-female (the styles of neuter and female florets are not different)

homolepis -is -e uniformly covered with scales

hondensis -is -e from Hondo Island, Japan

Honkenya for G.A. Honkeny (1724–1805), German botanist

hookerae for Lady Hooker (d. 1872), wife of Sir W.J. Hooker

Hookeria, hookeri, hookerianus -a -um for either Sir W.J. Hooker or his son Sir Joseph D. Hooker, both directors of Kew

horarius -a -um lasting for one hour (the expended petals of *Cistus*)

Hordelymus Barley-lime-grass

Hordeum Latin name for barley

horizontalis -is -e flat on the ground, spreading horizontally

horminoides clary-like, resembling *Horminium*

Horminum Exciter (the Greek name for sage used as an aphrodisiac)

hormo- chain-, necklace-

Hornungia for E.G. Hornung (1795–1862), German writer

horologicus -a -um with flowers that open and close at set times of day

horridus -a -um very thorny, rough, horridly armed

Hort for Arthur Hort (1864–1935), *Lychnis flos-jovis* and *Globularia meridionalis* cultivars bear his name

hort. signifying a plant name that is being used as the traditional gardeners' name – see *hortulanorum*

Hortensia A synonym for *Hydrangea*, for Hortense van Nassau

hortensis -is -e, hortorum cultivated, of the garden

hortulanus -a -um, hortulanorum of the gardener, of gardeners

hosmariensis -is -e from the neighbourhood of Beni Hosmar, Morocco

Hosta, hosteanus -a -um for Nicolas Tomas Host (1761–1834), physician

hostilis -is -e foreign

Hottonia for Peter Hotton (1648–1709), Swedish botanist

Houttuynia for Martin Houttuyn (1720–1794), Dutch naturalist

Howea (Howeia) from the Lord Howe Islands, East of Australia, or for Admiral Lord Richard Howe (1725–1799)

Hoya for Thomas Hoy, gardener for the Duke of Northumberland at Sion House

Hudsonia for William Hudson (1730–1793), English botanist

hugonis for Fr Hugh Scallon, collector in West China
humifusus -a -um spreading over the ground, trailing, sprawling
humilis -is -e, humilior low-growing, smaller than most of its kind
humuli- hop-, *Humulus*-like-
Humulus from the Slavic-German 'chmeli'
hungaricus -a -um from Hungary, Hungarian
hupehensis -is -e from Hupeh province, China
Hura from a South American vernacular name
Hutchinsia for Miss Hutchins (1785–1815), Irish cryptogamic botanist
Hyacinthoides Hyacinth-like
Hyacinthus Homer's name for the flower which sprang from the blood of υακινθος, or from an earlier Thraco-Pelasgian word, for the blue colour of water
hyacinthus -a -um, hyacinthinus -a -um dark purplish-blue, resembling *Hyacinthus*
hyalinus -a -um nearly transparent, crystal, hyaline
hybernalis -is -e, hybernus -a -um of winter
hybridus -a -um bastard, mongrel, cross-bred, hybrid
Hydrangea (Hortensia) Water-vessel (υδωρ–αγγος) the shape of the capsules
Hydrilla Water-serpent
hydro- water-, of water- (υδρο)
Hydrocharis Water-beauty
Hydrochloa Water-grass
Hydrocotyle Water-cup
hydrolapathum a name in Pliny, υδωρ–λαπαθον, for a water dock
hydropiper water pepper
hyemalis -is -e pertaining to winter, of winter (flowering)
hygro- moisture
hygrometricus -a -um responding to moisture level
Hygrophila Moisture-loving (υγρος moist) spiny plant of arid habitats, flowering in response to moisture
hylaeus -a -um of woods, of forests

hylo- forest, woodland (υλη)
Hylocereus Wood-cactus (climbing cactus)
hylophilus -a -um living in forests, wood-loving
hymen-, hymeno- membrane-, membranous- (υμην)
Hymenanthera Membranous-stamen (the membranous appendages of the anthers)
Hymenocallis Membranous-beauty (spider lily)
Hymenocardia Membranous-heart (the winged, heart-shaped fruits)
Hymenophyllum Membranous-leaf, delicate frond of the filmy fern
Hyoscyamus Hog-bean (a derogatory name, υοσκυαμος, by Dioscorides)
Hyoseris Pig-salad (swine's succory)
hyp-, hypo- under-, beneath- (υπο-)
hypanicus -a -um from the region of the Hypanis River, Sarmatia
hyparcticus -a -um beneath the Arctic
Hyparrhenia Male-beneath (the arrangement of the spikelets)
hyper- above-, over- (υπερ)
hyperboreus -a -um of the far North
hyperici- *Hypericum*-like-
Hypericum Above-pictures (ancient use over shrines to repel evil spirits)
Hyphaene Network (the fibres in the fruit wall)
hypnoides moss-like, resembling *Hypnum*
Hypnum Sleep (υπνος)
Hypochoeris a name, υποχοιρις, used by Theophrastus
hypochondriacus -a -um sombre, melancholy (colour)
hypochrysus -a -um golden underside, golden beneath
Hypodaphnis Inferior-laurel (the inferior ovary is unusual in the *Lauraceae*)
Hypoestes Below-house
hypogaeus -a -um underground, subterranean
hypoglossus -a -um beneath-a-tongue, sheathed-below
Hypolepis Under-scale (the protected sori)
hypoleucus -a -um whitish, pale

hypophegeus -a -um from beneath beech trees (but *Monotropa hypophegia* was parasitic on *Quercus*)
hypopithys, hypopitys growing under pine trees, πιτυζ
Hypsela High-one (υψηλος)
hyrcanus -a -um from the Caspian Sea area
hysginus -a -um dark reddish-pink
hyssopi- hyssop-like, resembling *Hyssopus*
Hyssopus from a Semitic word, ezob
hystri-, hystrix porcupine-like (the spiny corm of *Isoetes*)

iacinthus -a -um see *jacinthinus*
ianthinus -a -um, ianthus -a -um bluish-purple, violet-coloured
-ianus -a -um -pertaining to (a person), -'s
iaponicus -a -um see *japonicus*
-ias -much resembling
ibericus -a -um, ibiricus -a -um either from Spain and Portugal (Iberia) or from the Georgian Caucasus
iberideus -a -um from the Iberian peninsula
iberidi- *Iberis*-like
Iberis Iberia (Dioscorides' name, ιβηρις, for an Iberian cress-like plant)
-ibilis -is -e able-, capable of-
Icacina Icaco-like, resembling *Chrysobalanus icaco* (coco-plum)
-icans -becoming, -resembling
-icolus -a -um -of, -dwelling in
icos-, icosa- twenty-
icosandrus -a -um twenty-stamened
ictericus -a -um, icterinus -a -um yellowed, jaundiced
-icus -from (geographical names)
idaeus -a -um from Mt Ida in Crete, or Mt Ida in NW Turkey
-ides -resembling, -similar to, -like, -ειδης
Idesia for Eberhard Ysbrant Ides, Dutch explorer in China
idio- peculiar-, different- (ιδιος)
-idius -a -um -resembling

idoneus -a -um worthy, apt, suitable

Ifloga an anagram of *Filago*

ignescens, igneus -a -um kindling, fiery-red-and-yellow, glowing

ikariae from Ikaria

il-, im-, in- in-, into-, for-, contrary-, contrariwise-

Ilex the Latin name for the holm-oak (*Quercus ilex*)

ilici-, ilicinus -a -um holly-, *Ilex-*

-ilis -is -e -able, -having, -like, -resembling

illecebrosus -a -um alluring, enticing, charming

Illecebrum Charm (a name in Pliny)

Illicium Attractive (the fragrance)

-illimus -a -um -the best, -the most (superlative), -est

illinatus -a -um, illinitus -a -um smeared, smudged

-illius -a -um -lesser (a diminutive ending)

illustratus -a -um pictured, painted, as if painted upon

illustris -is -e brilliant, bright, clear

illyricus -a -um from Illyria, former Yugoslavia

ilvensis -is -e from the Isle of Elba, or the River Elbe

Ilysanthes Mud-flower

imbecillis -is -e, imbecillus -a -um feeble, weak

imberbis -is -e without hair, unbearded

imbricans, imbricatus -a -um overlapping like tiles (leaves, corolla, bracts, scales), imbricate

immaculatus -a -um unblemished, without spots, immaculate

immarginatus -a -um without a rim or border

immersus -a -um growing underwater

impari- unpaired-, unequal-

Impatiens Impatient (touch-sensitive fruits)

impeditus -a -um tangled, hard to penetrate, impeding

Imperata for the Italian botanist Imperato

imperator, imperatoria emperor, ruler, master

imperatricis for the Empress Marie Josephine Rose Tascher de la Pagerie (1763–1814)

imperforatus -a -um without perforations or apparent perforations *imperialis -is -e* very noble, imperial

implexus -a -um tangled, interlaced

impolitus -a -um dull, not shining, opaque

imponens deceptive

impressus -a -um sunken, impressed (e.g. leaf-veins), marked with slight depressions

impudicus -a -um lewd, shameless, impudent

in- not-, un-, en-, em-

inaequalis -is -e unequal-sided, unequal-sized (veins or other feature)

inaequidens with unequal teeth, not equally toothed

inapertus -a -um without an opening, closed, not opened

inarticulatus -a -um not jointed, indistinct (nodes)

inatophyllus -a -um thong-leaved

incanescens turning grey, becoming hoary

incanus -a -um quite grey, hoary-white

incarnatus -a -um flesh-coloured, carneus

Incarvillea for Pierre d'Incarville (1706–1757), correspondent of Bernard de Jussieu from China

incertus -a -um doubtful, uncertain

incisi-, *incisis -is -e*, *incisus -a -um* sharply and deeply cut into, incised

includens encompassed

inclusus -a -um not protruding, included (e.g. corolla longer than the style)

incomparabilis -is -e beyond compare, incomparable

incomptus -a -um unadorned, rough

inconspicuus -a -um small

incrassatus -a -um very thick, made stout (e.g. *Sempervivum* leaves)

incubaceus -a -um lying close upon the ground

incubus -a -um lying upon (when a lower distichous leaf overlaps the next on the dorsal side); Latin for a nightmare

incurvus -a -um, *incurvatus -a -um* inflexed, incurved

indicus -a -um from India or, loosely, from the Orient

Indigofera Indigo-bearer (source of blue dyes)

indivisus -a -um whole, undivided

induratus -a -um hard, indurate (usually of an outer surface)
induvialis -is -e, induviatus -a -um clothed with dead remnants (of leaves or other structure)
inebrians able to intoxicate, inebriating
inermis -is -e without spines or thorns, unarmed
-ineus -a -um -ish, -like
inexpectans not expected (found where not expected)
infarctus -a -um stuffed into, turgid
infaustus -a -um unfortunate
infectorius -a -um, infectoris dyed, used for dying, of the dyers
infestus -a -um troublesome, hostile, dangerous
infirmus -a -um weak, feeble
inflatus -a -um swollen, inflated
inflexus -a -um bent or curved abruptly inwards, inflexed
infortunatus -a -um unfortunate (poisonous)
infosus -a -um deeply sunken, buried
infra- below-
infractus -a -um curved inwards
infundibuliformis -is -e trumpet-shaped, funnel-shaped
ingens huge, enormous
innatus -a -um natural, inborn, innate
innominatus -a -um not named, unnamed
innoxius -a -um without prickles, harmless
inodorus -a -um without smell, scentless
inominatus -a -um unlucky, inauspicious
inophyllus -a -um fibrous-leaved, with fine thread-like veins
inopinatus -a -um, inopinus -a -um surprising, unexpected
inops deficient, poor, weak
inornatus -a -um without ornament, unadorned
inquilinus -a -um, inquillinus -a -um introduced
inquinans turning brown, staining, discolouring
inscriptus -a -um as though written upon, inscribed
insectifer -era -erum bearing insects (mimetic fly orchid)
insectivorus -a -um insect-eating

insertus -a -um inserted (the scattered inflorescences)

insignis -is -e remarkable, decorative, striking

insiticius -a -um, insititius -a -um, insitivus -a -um grafted

insubricus -a -um from the Lapontine Alps (*Insubria*) between Lake Maggiore and Lake Lucerne

insulanus -a -um, insularis -is -e growing on islands, insular

intactus -a -um unopened, untouched (the flowers)

integer -era -erum, integerrimus -a -um, integri- undivided, entire, intact, whole

integrifolius -a -um with entire leaves

inter- between-

interjectus -a -um intermediate in form, interposed (between two other species)

intermedius -a -um between extremes, intermediate

interruptus -a -um with scattered leaves or flowers

intertextus -a -um interwoven

intonsus -a -um bearded, unshaven, long-haired, leafy

intortus -a -um curled, twisted

intra-, intro- within-, inside-

intricatus -a -um entangled

introrsus -a -um facing inwards, turned towards the axis, introrse

intumescens swollen

intybus from a name in Virgil for wild chicory or endive

Inula a name in Pliny for *Inula helenium*, elecampane

inuncans covered with hooked hairs or glochidia

inunctus -a -um anointed

inundatus -a -um of marshes or places which flood periodically, flooded

-inus -a -um -ish, -like, -resembling, -from

invenustus -a -um lacking charm, unattractive

inversus -a -um turned over, inverted

involucratus -a -um surrounded with bracts, involucrate, with an involucre (the flowers)

involutus -a -um obscured, rolled inwards, involute

Iochroma Violet-colour (flower colour)
iodes violet-like, resembling *Viola*, ιον
iodinus -a -um violet-coloured
ioensis -is -e from Iowa, USA
ion-, *iono-* violet- (stock- or wallflower-)
-ion -occurring
ionantherus -a -um, *ionanthes* violet-flowered
ionanthus -a -um with violet-coloured flowers
ionenis -is -e from Iowa, USA
ionicus -a -um from the Ionian Islands, Greece
Ionopsis Violet-looking (violet cress)
ionosmus -a -um violet-scented
ipecacuanha a vernacular name for the drug producing *Cephaelis ipecacuanha*
Ipheion a name, ηθειον, used by Theophrastus
Ipomaea Worm-resembling (ιπς–ομοιος), the twining stems
Iresine Woolly (ειρος)
iricus -a -um from Ireland, Irish
iridescens iridescent
iridi- Iris-like
irio an ancient Latin name for a cruciferous plant
Iris the mythological name of the messenger of the Gods of the rainbow
irrigatus -a -um of wet places, flooded
irriguus -a -um watered (has clammy hairs)
irritans causing irritation
irroratus -a -um bedewed, dewy, *irroro*, to bedew
isabellinus -a -um drab-yellowish, tawny
isandrus -a -um equal-stamened, with equal stamens
Isatis the name, ισατις, used by Hippocrates for woad
Ischaemum Blood-stopper (a name, ισχω–αιμα, in Pliny for its styptic property)
-iscus -a -um -lesser (diminutive ending)

islandicus -a -um from Iceland, Icelandic
Isnardia for A.T.D. d'Isnard of Paris (1663–1743)
iso- equal- (ισος)
Isoetes Equal-to-a-year (green throughout)
Isolepis Equal-scales (the glumes)
Isoloma Equal-border (the equal lobes of the perianth)
Isolona Equal-petals (the equal petals)
Isotoma Equal-division (the equal corolla segments)
-issimus -a -um -est, -the best, -the most (superlative)
istriacus -a -um from Istria, former Yugoslavia
-ium -lesser (diminutive ending)
italicus -a -um from Italy, Italian
Itea from a Greek name for a willow, ιτεα
iteophyllus -a -um willow-leaved
-ites, -itis -closely resembling, -very much like, -ιτης
Iva an old name applied to various fragrant plants, used by Rufinus
ivorensis -is -e from the Ivory Coast, West Africa
Ixia Bird-lime (Theophrastus' name, ιξος, refers to the clammy sap)
Ixiolirion *Ixia*-lily (the superficial resemblance)
ixocarpus -a -um sticky-fruited
Ixora the name of a Malabar deity, Iswar

Jacaranda from the Tupi Guarani name, jakara'nda, for *J. cuspidifolia*
Jacea from the Spanish name for knapweed
jacinthinus -a -um reddish-orange coloured (*iacuntus*, relates to *Hyacinthus*)
jackmanii for G. Jackman, plant breeder of Woking
jacobaeus -a -um either for St James (Jacobus) or from Iago Island, Cape Verde
Jacobinia from Jacobina, Brazil
jalapa from Jalapa, Veracruz (*Mirabilis jalapa* false jalap); true purgative jalap is derived from *Ipomoea purga* (*Exogonium purga*)

jambolana from a Hindu name, jambosa, for *Eugenia jambolana*
jambos from a Malaysian name for rose-apple (*Eugenia jambos*)
Jamesia for Edwin James (1797–1861), American botanist
januensis -is -e from Genoa, N Italy, Genoan
japonicus -a -um (iaponicus -a -um) from Japan, Japanese
Jasione Healer (from a Greek name, ιασιονε, for *Convolvulus*)
jasminoides jasmine-like, resembling *Jasminum*
Jasminum latinized from the Persian name, yasmin
jaspidius -a -um, iaspidius jasper-like, striped or finely spotted in many colours
Jatropha Physician's-food (medicinal use)
javanicus -a -um from Java, Javanese
Jeffersonia for Thomas Jefferson (1743–1826), American President who strove to end slavery
jejunifolius -a -um insignificant-leaved
jejunus -a -um barren, poor, meagre, small
jezoensis -is -e from Jezo (Yezo), Hokkaido, Japan
jocundus -a -um see *jucundus*
jonquilla from the Spanish name for *Narcissus jonquilla*
jonquilleus -a -um the bright yellow of *Narcissus odorus*
johannis -is -e from Port St John, South Africa
Jovibarba Jupiter's-beard (the fringed petals)
juanensis -is -e from Genoa, N Italy, Genoan
Jubaea for King Juba of Numidia
jubatus -a -um maned (crested with awns)
jucundus -a -um pleasing, delightful
judaicus -a -um of Judaea, Jewish
judenbergensis -is -e from the Judenburg Mountains, Austria
jugalis -is -e, jugatus -a -um joined together, yoked
juglandi- *Juglans*-like-
Juglans Jupiter's-acorn (in Pliny – *glans Jovis*)
jugosus -a -um hilly, ridged
-jugus -a -um -yoked, -paired
jujuba from an Arabic name, jujube, for *Zizyphus jujuba*

juliae for Julia Mlokosewitsch who, in about 1900, discovered *Primula juliae*

julibrissin from the Persian name for *Acacia julibrissin*

julibrissius -a -um silken

juliformis -is -e downy

julii for Julius Derenberg of Hamburg, succulent grower

junceus -a -um, juncei-, junci- rush-like, resembling *Juncus*

Juncus Binder (*iungo*), classical Latin name refers to use for weaving and basketry

juniperinus -a -um bluish-brown, juniper-like, resembling *Juniperus* or its berry colour

Juniperus the ancient Latin name

junonia for the Greek goddess Juno

juranus -a -um from the Jura Mountains, France

Jussieua (Jussiaea) for Bernard de Jussieu (1699–1777), who made a major contribution to establishing the concept of the taxonomic species

Justicia for James Justice, Scottish gardener

juvenalis -is -e youthful

kaempferi for Engelbert Kaempfer (1651–1715), German physician and botanist

kaido a Japanese name

kaki from the Japanese name, kaki-no-ki, for persimmon (*Diospyros kaki*)

Kalanchoe from a Chinese name

kali, kali- either from the Persian for a carpet, or a reference to the ashes of saltworts being alkaline (alkali); cognate with Kalium (potassium)

Kalmia for Peter Kalm (1716–1779), a highly reputed student of Linnaeus

kalo- beautiful-, καλος–

Kalopanax Beautiful-*Panax*

kamtschaticus -a -um from the Kamchatka Peninsula, E Siberia

kansuensis -is -e from Kansu province, China
karwinskii, karwinskianus -a -um for Wilhelm Friederich Karwinsky von Karwin (1780–1855), plant collector in Brazil
katangensis -is -e from Katanga (Shaba)
katherinae for Mrs Katherine Saunderson, who collected plants in Natal
keleticus -a -um charming
Kentranthus Spur-flower (see *Centranthus*)
kermesinus -a -um carmine-coloured, carmine
Kerria for William Kerr (d. 1814), collector of Chinese plants at Kew
Keteleeria for J.B. Keteleer, French nurseryman
kewensis -is -e of Kew Gardens
khasianus -a -um from the Khasia Hills, Assam, N India
Kickxia for J.J. Kickx (1775–1831), Belgian cryptogamic botanist
Kigelia from the native Mozambique name for the sausage tree
kingdonii for Capt. F. Kingdon-Ward
Kirengeshoma from the Japanese, ki- (yellow) -rengeshoma (*Anemopsis macrophylla*)
kirro- citron-coloured-
kisso- ivy-, ivy-like-
Kitaibela (Kitaibelia) for Paul Kitaibel (1757–1857), botanist at Pécs, Hungry
kiusianus -a -um from Kyushiu, one of the major islands forming S Japan
Kleinia for J. Th. Klein (1685–1759), German botanist
Knautia for Christian Knaut (1654–1716), German botanist
Kniphofia for Johann H. Kniphof (1704–1763), German botanist
Kobresia (Cobresia) for Carl von Cobres (1747–1823), Austrian botanist
kobus from a Japanese name, kobushi
Kochia for Wilhelm Daniel Joseph Koch
Koeleria for L. Koeler, German botanist
Koelreuteria for Joseph Gottlieb Koelreuter (1733–1806), Professor of Natural History, Karlsruhe

Koeningia (Koenigia, Koeniga) for J.G. König (1728–1785), student of Linnaeus, botanist in India

Kohlrauschia for F.H. Kohlrausch, assiduous German lady botanist

Kolkwitzia for Richard Kolkwitz, Professor of Botany, Berlin

kolomicta a vernacular name from Amur, E Siberia, for *Actinidia kolomicta*

koreanus -a -um, koraiensis -is -e from Korea, Korean

kotschianus -a -um for Theodor Kotschy

kousa a Japanese name for *Cornus kousa*

kurroo from a Himalayan name for *Gentiana kurroo*

labdanus -a -um see *ladanum*

-labellus -a -um -lipped, -with a small lip (*labrum*, a lip)

labiatus -a -um lip-shaped, lipped, labiate

labilis -is -e unstable, labile

labiosus -a -um conspicuously lipped

lablab from a Hindu name for hyacinth bean, *Dolichos lablab*

Laburnum an ancient Latin name used by Pliny

lac-, lacto- milky- (*lac* milk)

-lacca, lacco- -resin, varnished-

lacciferus -a -um producing a milky juice

Laccodiscus Varnished-disc (the shining floral disc)

lacer, lacerus -era -erum, laceratus -a -um torn into a fringe, as if finely cut into

lacertinus -a -um lizard-tailed (the common garden lizard is *Lacerta vivipara*)

Lachenalia for Werner de La Chenal (de Lachenal) (1763–1800), Swiss botanist

lachno- downy-, woolly-

Lachnostoma Woolly-mouth (the throat of the corolla is bearded)

lachnopus -a -um woolly-stemmed, downy-stalked

laciniatus -a -um, laciniosus -a -um jagged, fringed, slashed (see Fig. 4(*f*))

lacistophyllus -a -um having torn leaves

lacrimans (lachrymans) causing tears, weeping branching habit (*lacrima* tear)
lacryma-jobi Job's-tears (the shape and colour of fruit)
lactescens having lac, or milky sap
lacteus -a -um, *lact-*, *lacti-* milk-coloured, milky-white
lactifer -era -erum producing a milky juice
Lactuca the Latin name (has a milky juice)
lacunosus -a -um with gaps, furrows, pits or deep holes
lacuster, *lacustris -is -e* of lakes or ponds (*lacus* lake)
ladanifer -era -erum bearing *ladanum*, ληδανον gum (the resin called myrrh)
ladysmithensis-is-e from Ladysmith, South Africa
Laelia after one of the Vestal Virgins
laetevirens bright-green
laeti-, *laetis -is -e*, *laetus -a -um* pleasing, vivid, bright
laevi-, *laevigatus -a -um*, *laevis -is -e* polished, not rough, smooth (*levis* smooth)
laevo- to the left-
lag-, *lago-* hare's-
lagaro-, *lagaros-* lanky-, long-, thin-, narrow-, λαγαρος
Lagarosiphum Narrow-tube
lagen-, *lagenae-*, *lageni-* bottle-
lagenaeflorus -a -um with flask-shaped flowers
Lagenaria Flask (the bottle-gourd fruit of *Lagenaria siceraria*)
lagenarius of a bottle or flask
Lagerstroemia for Magnus von Lagerstrom of Goteborg (1696–1759), friend of Linnaeus
lagopinus -a -um hare's-foot-like
lagopodus -a -um hare's foot, λαγωπους
lagopus hare's foot (an old generic name)
Laguncularia Small-bottle (the fruit)
lagunensis -is -e from Laguna, Luzon, Philippines
Lagurus Hare's-tail (the inflorescence)
lakka from a vernacular name for the palm *Cyrtostachys lakka*

Lamarckia (Lamarkia) for Jean Baptiste Antoine Pierre Monnet de Lamarck (1744–1829), French evolutionist

lamellatus -a -um layered, lamellate (diminutive of *lamellus* sheet)

lamii- deadnettle-like, resembling *Lamium*

laminatus -a -um laminated

Lamiopsis Looking-like-*Lamium*

Lamium Gullet (the name in Pliny refers to the gaping mouth of the corolla)

lampas lamp-like, bright

lampr-, lampro- shining-, glossy-, λαμπρος

Lampranthus Shining-flower

lanatus -a -um woolly

lancastriensis -is -e from Lancashire, Lancastrian

lanceolatus -a -um, lanci- narrowed and tapered at both ends, lanceolate

lancerottensis -is -e from Lanzarote, Canary Isles

lanceus -a -um spear-shaped

landra from the Latin name for a radish

langleyensis -is -e from Veitch's Langley Nursery, England

lani-, laniger -era -erum, lanosus -a -um, lanuginosus -a -um softly-hairy, woolly or cottony

lanigerus -a -um woolly

Lantana an old Latin name for *Viburnum*

lantanoides resembling *Lantana*

lanthanum inconspicuous (λανθανειν escape notice)

lanugo soft-haired (*lana* wool)

Lapageria for Marie Josephine Rose Tascher de la Pagerie (1763–1814), Napoleon's Empress Josephine, avid collector of roses at Malmaison

lapathi- sorrel-like-, dock-like-, λαπαθον

Lapathum Adanson's use of the Latin name for sorrel

Lapeirousia (Lapeyrousia) for J.F.G. de La Peyrouse (1741–1788), French circumnavigator

lapidius -a -um hard, stony

lappa Latin name for bur-fruits (e.g. goosegrass and burdock, *Arctium lappa*)

lappaceus -a -um bearing buds, bud-like, burdock-like

lapponicus -a -um, lapponus -a -um from Lapland, of the Lapps

lappulus -a -um with small burs (the nutlets)

Lapsana (Lampsana) Purge (Dioscorides' name for a salad plant)

larici-, laricinus -a -um larch-like, resembling *Larix*

laricinifolius -a -um larch-leaved

laricio the Italian name for several pines

Larix Dioscorides' name for the larch

lascivius -a -um running wild, impudent

Laser a Latin name for several umbellifers

Laserpitium an ancient Latin name

lasi-, lasio- shaggy-, woolly-, λασιος-

lasianthus -a -um with shaggy flowers

lasiopcarpus -a -um having woolly fruits

lasiolaenus -a -um shaggy-cloaked, woolly-coated

Lasiopetalum Woolly-petals (the sepals are downy and petaloid)

Lastrea for C.J.L. de Lastre (1792–1859), French botanical writer

latebrosus -a -um of dark or shady places

lateralis -is -e, lateri- on the side, laterally-

latericius -a -um, lateritius -a -um brick-red

Lathraea Hidden (until flowering, inconspicuous root parasites), λαθραιος

lathyris the name for a kind of spurge (*Euphorbia lathyris*)

Lathyrus the ancient name, λαθυρος, for the chickling pea (*Lathyrus sativus*) used by Theophrastus

lati-, latisi-, latus -a -um broad-, wide-

latici- latex-, juice-

latifrons with broad fronds

latipes broad-stalked, thick-stemmed

latiusculus -a -um somewhat broad

latobrigorum of the Rhinelands

laudatus -a -um praised, worthy, lauded

laureola Italian name for *Daphne laureola*, from its use in garlands
lauri- laurel-, *Laurus*-like-
lauricatus -a -um wreathed, resembling laurel or bay
laurinus -a -um laurel-like
laurocerasus laurel-cherry (cherry-laurel)
Laurus the Latin name for laurel or bay
Laurustinus Laurel-like-*Tinus*
lautus -a -um washed
Lavandula To-wash (its use in the cleansing process)
lavandulae- lavender-, *Lavandula*-
Lavatera for the brothers Lavater, 18th century Swiss naturalists
lavateroides *Lavatera*-like
Lawsonia for Dr Isaac Lawson, botanical traveller (henna plant)
lawsonianus -a -um for P. Lawson (d. 1820), Edinburgh nurseryman
laxi-, *laxus -a -um* open, loose, not crowded, distant, lax
lazicus -a -um from NE Turkey (Lazistan)
lecano- basin-
Lecanodiscus Basin-disc (the concave floral disc)
Lecythis Oil-jar (ληκυθος) the shape of the fruit with its lid
Ledebouria for Carl Friedrich von Ledebour (1785–1851), student of the Russian flora
Ledum an ancient Greek name, ληδανον, for the ladanum-resin producing *Cistus ladaniferus*
Leersia for J.D. Leers (1727–1774), German botanist
legionensis -is -e from Leon, Spain
Legousia etymology uncertain
leio- smooth- (λειος)
Leiophyllum Smooth-leaf
Lemna Theophrastus' name, λεμνα, for a water-plant
lemniscatus -a -um beribboned (the Roman victor had ribbons, *lemnisci*, from his crown)
lendiger -era -erum nit-bearing (the appearance of the spikelets)
Lens the classical name for the lentil
lenti- spotted-, freckled-

Lentibularia usually regarded as referring to the lentil (lens)-shaped bladders

lenticularis -is -e lens-shaped, bi-convex

lenticulatus -a -um with conspicuous lenticels on the bark, lenticulate

lentiformis -is -e lens-shaped, bi-convex

lentiginosus -a -um freckled, mottled

lentiscus Latin name for the mastic tree, *Pistacia lentiscus*

lentus -a -um tough, pliable

leo-, leon- lion-, λεων

leodensis -is -e from Liège, Belgium

leonensis -is -e from Sierra Leone, West Africa

leoninus -a -um tawny-coloured like a lion

leonis -is -e toothed or coloured like a lion

Leonotis Lion's-ear

leonto- lion's-

Leontodon Lion's-tooth

Leontopodium Lion's-foot

Leonurus Lion's-tail

leopardinus -a -um conspicuously spotted

Lepidium Little-scale (Dioscorides' name, λεπιδιον, for a cress refers to the fruit)

lepido-, lepiro- flaky-, scaly-, λεπις (the scales may be minute as on butterflies' and moths' wings)

lepidus -a -um neat, charming

Lepidobotrys Scale-cluster (the flowers emerge from strobilus-like groups of subtending bracts)

lepidocaulon with a scaly stem

lepidopteris scale-winged

Lepidotis Scaly

lepidotus -a -um scurfy, scaly, lepidote

lepidus -a -um neat, elegant, graceful

-lepis -scaly, -scaled, -λεπις

Lepiurus Scale-tail (the inflorescence of sea hard grass, cf. *Pholiurus*)

leporinus -a -um hare-like

leprosus -a -um scurfy, leprosied (λεπρα)
lept-, lepta-, lepto- slender-, weak-, thin-, small-, delicate-, λεπτος-
Leptactinia Slender-rayed (the circlet of fine corolla lobes)
leptochilus -a -um with a slender lip
leptoclados with slender shoots
Leptodermis Thin-skin (the inner fruit-wall)
Leptogramma Slender-lined (the sori)
Leptonychia Slender-clawed (the staminodes)
leptophis -is -e slender
leptophyllus -a -um slender-leaved
Leptospermum Narrow-seed (slender-seeded)
Lepturus Hare's-tail
Lespedeza for V.M. de Lespedez, Spanish politician in Florida
leuc-, leuco- white-, λευκο-
Leucadendron White-tree
Leucanthemum White-flower (Dioscorides' name), *Chrysanthemum*
Leucanthemella Little-white-flower, *Chrysanthemum*
leuce a name for the white poplar
Leucobryum White-*Bryum* (the greyish-white appearance)
leucochroa white-coloured, pale
Leucojum White-violet (Hippocrates' name, λευκο–ιον, for a snowflake)
Leucorchis White-orchid
Leucothoe an ancient Greek name, Leucothoe was daughter of King Orchanus of Babylon
levigatus -a -um smooth, polished
levis -is -e smooth, not rough
Levisticum Alleviator (the Latin equivalent of the Dioscorides' Greek name λιγυστιχος)
Lewisia for Captain Meriwether Lewis (1774–1809) of the trans-American expedition
Leycesteria for William Leycester, judge and horticulturalist in Bengal *c.* 1820
lhasicus -a -um from Lhasa, Tibet

Liatris derivation uncertain

libanensis -is -e, libanoticus -a -um from Mt Lebanon, Syria

libani from the Lebanon, Lebanese

libanotis -is -e from Mt Lebanon or of incense, λιβανωτις

libericus -a -um from Liberia, West Africa

liber unrestricted, undisturbed

libero- bark- (a characteristic)

liberoruber with red bark

Libertia for Marie A. Libert (1782–1865), Belgian writer on Hepatics

Libocedrus Crying-cedar (the resin exudate of the incense cedar)

liburnicus -a -um from Croatia (*Liburnia*) on the Adriatic

libycus -a -um from Libya, Libyan

lignescens turning woody

ligni- woody-, wood-, of woods-

lignosus -a -um woody

lignum-vitae wood-of-life (the remarkably durable timber of *Guaiacum officinale*)

ligtu from a Chilean name for St Martin's flower

Ligularia Strap, *ligula* (the shape of the ray florets)

ligularis -is -e strap-shaped, ligule-like

ligulatus -a -um with a ligule, with a membranous projection, ligulate

Ligusticum, ligusticus -a -um Dioscorides' name, λιγυστικος, for a plant from Liguria, NE Italy

ligustrinus -a -um privet-like, resembling *Ligustrum*

Ligustrum Binder (a name used in Virgil)

lilacinus -a -um lilac-coloured, lilac-like

lili-, lilii- lily-

liliaceus -a -um lily-like, resembling *Lilium*

liliago silvery

Lilium the name in Virgil

lilliputianus -a -um of very small growth, Lilliputian

limaci- slug-

limaeus -a -um of stagnant waters
limbatus -a -um bordered, with a margin or fringe
-limbus -a -um -bordered, -fringed
limbo- border-, margin-, *limbus*
limensis -is -e from Lima, Peru
limicolus -a -um living in mud
Limnanthemum Pond-flower (spreads over surface)
Limnanthes Pond-flower
limn-, limno- marsh-, pool-, pond-, λιμνη
limnophilus -a -um marsh-loving
limon the Persian name for *Citrus* fruits
Limonium Meadow-plant (Dioscorides' name, λειμωνιον, for a meadow plant)
Limosella Muddy
limosus -a -um muddy, slimy, living on mud (*limus* mud)
lin-, linarii-, lini- flax-
linaceus -a -um flax-like, resembling *Linum*
Linaria Flax-like (the leaf similarity of some species)
Lindleyella for Dr John Lindley (1799–1865), Secretary and saviour of the Royal Botanic Gardens, Kew
linearis -is -e narrow and parallel-sided (usually the leaves), linear
lineatus -a -um marked with lines (usually parallel and coloured), striped
linicola dweller of flax fields
lingua, linguae-, lingui- tongue-shaped-, lingulate- (some structure or part)
-linguatus -a -um, -lingus -a -um -tongued
lingularis -is -e, lingulatus -a -um, linguus -a -um tongue-shaped (*Linguus* was a name in Pliny)
linicolus -a -um of flax-fields
linitus -a -um smeared
Linnaea by Gronovius, at request of Carl Linnaeus, for its lowly, insignificant and transient nature
linnaeanus -a -um, linnaei for Carl Linnaeus

linoides flax-like, resembling *Linum*

linosyris yellow-flax, an old generic name by l'Obel

Linum the ancient Latin name for flax

lio- smooth- (λειος)

liolaenus -a -um smooth-cloaked, glabrous

Liparis Greasy (the leaf-texture, λιπαρος)

lipo- greasy- (λιπος)

Lippia for A. Lippi (b. 1678), French/Italian naturalist

Liquidambar Liquid-amber (the fragrant resin from the bark of sweet gum, *Liquidambar styraciflua*)

liratus -a -um ridged (*lira*, a ridge)

lirelli- with a central furrow-

lirio- lily-white-

Liriodendron Lily-tree (the showy flowers of the tulip tree)

Liriope for one of the Nymphs of Greek mythology, the tail-end of a hood

liss-, lisso- smooth-

Lissochilus Smooth-lip (of the corolla)

Listera for Dr M. Lister (1638–1712), pioneer palaeontologist

Litchi from the Chinese vernacular name

literatus -a -um with the appearance of being written upon

litho- stone- (λιθος)

Lithocarpus Stone-fruit (the hard shell of *Lithocarpus javensis*)

lithophilus -a -um living in stony places, stone-loving

Lithops Stone-like (the mimetic appearance of stone-cacti)

Lithospermum Stone-seed (Dioscorides' name, λιθοσπερμον, for the glistening, whitish nutlets)

lithuanicus -a -um from Lithuania, Lithuanian

litigiosus -a -um disputed, contentious

litoralis -is -e, littoralis -is -e, littorius -a -um growing by the shore, of the sea-shore

Littonia for Samuel Litton

Littorella Shore (the habitat)

lituus -a -um forked and with the ends turned outwards

lividus -a -um lead-coloured, bluish-grey, leaden

Livistonia for Patrick Murray, Lord Livingstone, whose garden formed the nucleus of the Edinburgh Royal Botanic Garden

Lizei for the Lizé Frères of Nantes, France

llano- of treeless savannah-

Lloydia for Curtis G. Lloyd (1859–1926), American botanist

Loasa from a South American vernacular name

lobatus -a -um, lobi-, lobus -a -um with lobes, lobed (see Fig. 4(*d*))

lobbii, lobbianus -a -um for the brothers Thomas and William Lobb

Lobelia for Matthias de l'Obel (1538–1616), renaissance pioneer of botany and herbalist to James I of England

lobiferus -a -um having lobes

-lobium, -lobion -pod, -podded

Lobivia an anagram of Bolivia, provenance of the genus

Lobularia Small-pod (*lobulus*a small lobe)

-lobus -a -um -lobed (λοβος)

lochabrensis -is -e from Lochaber, Scotland

lochmius -a -um coppice-dweller, of thickets

-locularis -is -e -celled (usually the ovary)

locusta spikeleted, old generic name for *Valerianella locusta* (crayfish or locust)

loganobaccus -a -um loganberry, after its developer Judge J.H. Logan of California

Loiseleuria for Jean Louis August Loiseleur-Deslongchamps (1774–1849), French botanist

loliaceus -a -um resembling *Lolium*

Lolium a name in Virgil for a weed grass

loma- Peruvian grass-steppe-

-loma -fringe, -border

Lomaria Bordered (the marginal sori, λομα)

Lomariopsis *Lomaria*-like

Lomatia Fringed (the seeds are wing-bordered)

lomentiferus -a -um bearing constricted pods that break up into one-seeded portions *lonchitis -is -e*, *loncho-* spear-shaped, lance-shaped (a name, λονχιτις, used by Dioscorides for a fern)

Lonchocarpus Lance-fruit, λογχη–καρπος (the flat, indehiscent pods)

lonchophyllus -a -um with spear-like leaves

longe-, *longi-*, *longus -a -um* long-, elongated-

longipes long-stalked

Lonicera for Adam Lonitzer (1528–1586), German physician and botanist

lophanthus -a -um with crested flowers

Lophhira Crested (one of the sepals enlarges to a wing which aids fruit dispersal)

lopho- crest-, crested- (λοφια)

lophogonus -a -um crested-angular, with crested angles (as on a stem or fruit)

lophophilus -a -um living on hills, hill-loving

Lophophora Crest-bearer (has tufts of glochidiate hairs)

lora-, *loratus -a -um*, *lori-*, *loro-* strap-shaped, λωρον

Loranthus Strap-flower (the shape of the 'petals')

loricatus -a -um with a hard protective outer layer, clothed in mail

loriceus -a -um armoured, with a breast-plate

lorifolius -a -um with long narrow leaves, strap-leaved

Loroglossum Strap-tongue (the elongate lip)

loti-, *lotoides* trefoil-like, resembling *Lotus*

Lotus the ancient Greek name for various leguminous plants

louisianus -a -um from Louisiana, USA

loxo- oblique-

Loxogramma Oblique-lettered (the sori)

lubricus -a -um smooth, slippery

lucens, *lucidus -a -um* glittering, clear, shining

luciae for Madame Lucie Savatier

lucianus -a -um from St Lucia, West Indies

luciliae for Lucile Boissier

luconianus -a -um from Luzon, Philippines

lucorum of woodland or woods
Luculia from a Nepalese vernacular name
ludens of games, sportive
ludovicianus -a -um from Louisiana, USA
Ludwigia for C.G. Ludwig (1709–1773), German botanist
Luffa from the Arabic name, louff, for *Luffa cylindrica*
lugdunensis -is -e from Lyons, France
lugens mourning, downcast
luma from a Chilean vernacular name for *Myrtus luma*
Lunaria Moon (the shape and colour of the septum (or replum) of the fruit of honesty)
lunatus -a -um half-moon-shaped, lunate
lunulatus -a -um crescent-moon-shaped
lupicidius -a -um wolf's-bane
lupinellus -a -um like a small *Lupinus*
Lupinus the ancient Latin name for the white lupin
lupuli-, lupulinus -a -um hop-like, with the rampant habit of *Humulus lupulus*
Lupulus Wolf, in reference to its straggling habit on other plants (the ancient Latin name for hop was *Lupus salictarius*–willow wolf)
luridus -a -um sallow, dingy yellow or brown, wan, lurid
Luronium Rafinesque's name for a water plantain
lusitanicus -a -um from Portugal (Lusitania), Portuguese
lutarius -a -um of muddy places, living on mud
luteo-, luteus -a -um yellow
luteolus -a -um yellowish
lutescens turning yellow
lutetianus -a -um from Paris (Lutetia), Parisian
lutra otter
luxatus -a -um dislocated
luxurians rank, exuberant, luxuriant, of rapid growth
luzuli- *Luzula*-like
Luzula an ancient name of obscure meaning
Lycaste for Lycaste, daughter of King Priam of Troy

Lychnis Lamp (the hairy leaves were used as wicks for oil lamps, λυχνις)
lychnitis from a name in Pliny meaning of lamps
lychno-, lychnoides *Lychnis*-like
Lycium the ancient Greek name, λυκιον, for a thorn tree from Lycia
lycius -a -um from Lycia, SW Turkey
lyco- wolf- (usually implying inferior or wild) (λυκος)
Lycocarpus Wolf-fruit (clawed at the upper end)
lycoctonus -a -um wolf-murder (poisonous wolf's-bane, *Aconitum lycoctonum*)
lycoides box-thorn-like, similar to *Lycium*
Lycopersicum (on) Wolf-peach (tomato)
Lycopodium Wolf's-foot (clubmoss)
Lycopsis Wolf-like (Dioscorides' derogatory name, λυκοψις)
lycopsoides resembling *Lycopsis*
Lycopus Wolf's-foot
Lycoris for Lycoris the actress, and Marc Antony's mistress
lydius -a -um from Lydia, SW Turkey
Lygodium Willow-like (the climbing fern's stems)
lynceus -a -um lynx-like? (Lynceus was a keen sighted Argonaut)
lyratus -a -um lyre-shaped (rounded above with small lobes below–usually of leaves)
lysi-, lysio- loose-, loosening- (λυσις)
Lysichiton (um) Loose-cloak, λυσις–χιτων (the open, deciduous spathe)
Lysimachia Ending-strife, λυσιμαχος, named after theThracian king Lysimachus
Lythrum Black-blood (Dioscorides' name, λυθρον, may refer to the flower colour of some species)

Maakia for Richard Maack (1825–1886), Russian naturalist
Macaranga from the Malayan vernacular name, umbrella tree (the large leaves)
macedonicus -a -um from Macedonia, Macedonian

macellus -a -um rather meagre, poorish
macer -ra -rum meagre
macilentus -a -um thin, lean
Macleaya for Alexander Macleay (1767–1848), Secretary to the Linnaean Society of London
macr-, macro- big-, large-, long- (μακρος)
macrodus -a -um large-toothed
macromeris -is -e with large parts
macrorhizus -a -um large-rooted
macrosiphon large-tubular, long-tubed
macrurus -a -um (macrourus) long-tailed
maculatus -a -um, maculosus -a -um, maculifer -era -erum spotted, blotched, bearing spots
maculi- spot-like-
madagascariensis -is -e from Madagascar, Madagascan
maderaspatanus -a -um, maderaspatensis -is -e from the Madras region of India
maderensis -is -e from Madeira, Madeiran
Madia from a Chilean name
madrensis -is -e from the Sierra Madre, northern Mexico
madritensis -is -e from Madrid, Spain
Maerua from an Arabic name, meru
maesiacus -a -um from the Bulgarian–Serbian region once called Maesia
Maesobotrya *Maesa*-like-fruited (similarity of the fruiting clusters)
magellanicus -a -um from the Straits of Magellan, South America
magellensis -is -e from Monte Majella, Italy
magni-, magno-, magnus -a -um large
magnificus -a -um great, eminent, distinguished, magnificent
magnifolius -a -um large-leaved
Magnolia for Pierre Magnol (1638–1715), director of Montpelier Botanic Garden
mahagoni mahogany, from a South American vernacular name for *Swietenia mahagoni*

mahaleb an Arabic vernacular name for *Prunus mahaleb*
Mahernia an anagram of *Hermannia*, a related genus
Mahonia for Bernard McMahon (1775–1816), American horticulturalist
mai-, maj- May- (*maius* May)
Maianthemum May-flower, a May-flowering lily
mairei for Edouard Maire
majalis -is -e (magalis) of the month of May (flowering time)
majesticus -a -um majestic
major -or -us larger, greater, bigger
malabaricus -a -um from the Malabar coast, S India
malaco-, malako-, malacoides soft, μαλακος, tender, weak, mucilaginous, mallow-like
malacophilus -a -um pollinated by snails, snail-loving
malacophyllus -a -um with soft or fleshy leaves
Malaxis Softening, μαλαξις (soft leaves)
Malcolmia (Malcomia) for William Malcolm, English horticulturalist c. 1798
maleolens of bad fragrance, stinking
maliformis -is -e apple-shaped
mallococcus -a -um downy-fruited
mallophorus -a -um wool-bearing
Mallotus Woolly (the fruits of some species)
Malope a name for mallow in Pliny
Malpighia for Marcello Malpighi (1628–1694), Italian naturalist
Malus the ancient Latin name for an apple tree
Malva Soft (the name in Pliny)
malvaceus -a -um mallow-like, resembling *Malva*
Malvaviscus Mallow-glue (Wax mallow)
malvinus -a -um mauve, mallow-like
mammaeformis -is -e, mammiformis -is -e shaped like a nipple
Mammea from a West Indian vernacular name

Mammillaria (Mamillaria) Nippled (conspicuous tubercles)
mammillaris -is -e, mamillarius -a -um, mammillatus -a -um having nipple-like structures, mammillate
mammosus -a -um full-breasted
mancus -a -um deficient, inferior
mandibularis -is -e jaw-like, having jaws
Mandragora a Greek name derived from a Syrian mandrake
mandschuricus -a -um, mandshuricus -a -um from Manchuria, Manchurian
Mangifera from the Hindu name for the mango fruit
manicatus -a -um with long sleeves, with a felty covering which can be stripped off, manicate
Manihot from the Brazilian name for cassava
manipuliflorus -a -um with few-flowered clusters
manipuranus -a -um from Manipur, India
mano- scanty-, μανος
manriqueorum for Manrique de Lara, of the Manriques
mantegazzianus -a -um for Paulo Mantegazzi (1831–1910), Italian traveller and anthropologist
Manzanilla from the Spanish, manzanita, for a small apple
Maranta for Bartolomea Maranti, Venetian botanist
Marattia for J.F. Maratti (d. 1777), Italian botanist, author of *De Floribus Filicum*
marcescens not putrefying, persisting, retaining dead leaves and/or flowers
marckii for Jean Baptiste Antoine Pierre Monnet de la Marck (Lamarck) (1744–1829); French pre-Darwinian evolutionist
margaritaceus -a -um, margaritus -a -um pearly, of pearls
margaritiferus -a -um bearing pearl
marginalis -is -e of the margins, margined
marginatus -a -um having a distinct margin (the leaves)
Margyricarpus Pearl-fruit (μαργαριτης), the white berry-like achenes

marianus -a -um of St Mary, from Maryland, USA, or from the Sierra Morena

mariesii for Charles Maries (1850–1902), English plant collector

marilandicus -a -um, marylandicus -a -um from the Maryland region, USA

marinus -a -um growing by or in the sea, marine

mariscus -a -um the name for a rush in Pliny

maritimus -a -um growing by the sea, maritime, of the sea

marjoranus -a -um derived from the Latin name, *margorana*, for sweet marjoram

marmelos a Portuguese vernacular name, marmelo, for marmalade

marmoratus -a -um, marmoreus -a -um with veins of colour, marbled

maroccanus -a -um from Morocco, NE Africa, Moroccan

Marrubium the old Latin name

Marsdenia for Willam Marsden (1754–1836), author of a history of Sumatra

Marsilea for Ludwig F. Marsigli (1658–1730), Italian patron of botany

marsupiflorus -a -um with purse-like flowers

martagon resembling a kind of Turkish turban

Martia, Martiusia for K.F.P. von Martius (1794–1868), German botanist in Brazil

martinicensis -is -e from Martinique

maru mastic

marus -a -um glowing

mas, maris, masculus -a -um bold, with stamens, male

masculinus -a -um male, staminate

massiliensis -is -e from Marseilles, France

mastichinus -a -um gummy, mastic-like, like the mastic exuded by *Pistacia lentiscus*

mastigophorus -a -um (producing gum, gum-bearing) whip-bearing

Matricaria Mother-care (former medicinal use in treatment of uterine infections)

matritensis -is -e from Madrid, Spain

matronalis -is -e of married women (the Roman matronal festival was held on March 1st)
matsudana for Sadahisa Matsudo (1857–1921), Japanese botanist
Matteuccia (Matteucia) for C. Matteucci (1800–1868), Italian physicist
Matthiola for Pierandrea A.G. Matthioli (1500–1577), Italian botanist
matutinalis -is -e, matutinus -a -um morning, of the morning, early
mauritanicus -a -um from Morocco or North Africa generally
mauritianus -a -um from the island of Mauritius, Indian Ocean
maurorum of the Moors, Moorish, of Mauritania
maurus -a -um from Morocco
maxillaris -is -e of jaws, resembling an insect's jaws
maximus -a -um largest, greatest
mays from the Mexican name for Indian corn
Mazus Nipple (μαζος) the shape of the corolla
meandriformis -is -e of winding form, much convoluted
meanthus -a -um small-flowered
-mecon -poppy, μηκων
Meconopsis Poppy-like
medeus -a -um remedial, healing, curing
medi-, medio-, medius -a -um middle-sized, between-, intermediate-
Medicago from a Persian name for a grass
medicus -a -um from Media (Iran), curative, medicinal
mediolanensis -is -e from Milan, Italy
mediopictus -a -um with a coloured stripe down the centre-line (of a leaf)
mediterraneus -a -um from the Mediterranean region, from well inland
medullaris -is -e, medullus -a -um pithy, soft-wooded
medullarius -a -um, medullosus -a -um with a large pith
mega-, megali-, megalo- big-, great-, large-, μεγαλη-
megacephalus -a -um large-headed (of composite inflorescences)
megalurus -a -um large-tailed

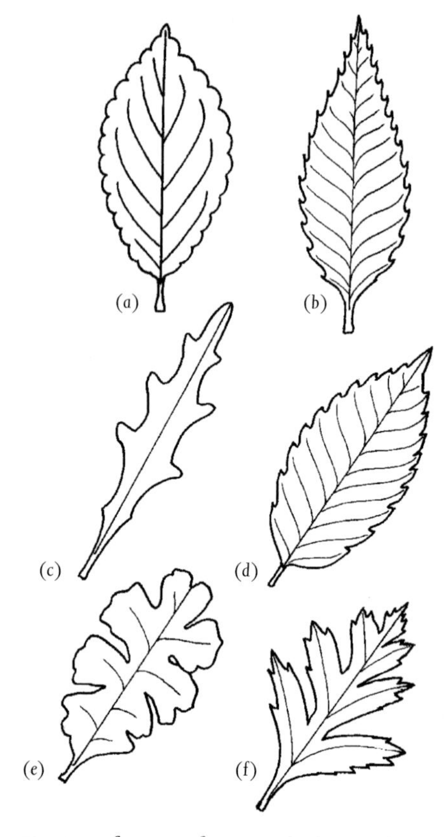

Fig. 4. Leaf-margin features which provide specific epithets.
(*a*) Crenate (scalloped as in *Ardisia crenata* Sims);
(*b*) dentate (toothed as in *Castanea dentata* Borkh.). This term has been used for a range of marginal tooth shapes;
(*c*) sinuate (wavy as in *Matthiola sinuata* (L.) R.Br.). This refers to 'in and out' waved margins, not 'up and down' or undulate waved margins;
(*d*) serrate (saw-toothed as in *Zelkova serrata* (Thunb.) Makino);
(*e*) lobate (lobed, as in *Quercus lobata* Nee);
(*f*) laciniate (cut into angular segments as in *Crataegus laciniata* Ucria).

megapotamicus -a -um of the big river, from the Rio Grande or River Amazon

megaseifolius -a -um *Megasea*-leaved (*Bergenia*-leaved)

meio- (meon-) fewer, less than-, μειων- (meiosis is the reduction division during spore formation)

meiophyllus -a -um with fewer leaves in each successive whorl

meiostemonus -a -um with fewer stamens

mela-, melan-, melano- black- (μελας, μελανος)

Melaleuca Black-and-white (the colours of the bark on trunk and branches)

Melampyrum (on) Black-wheat (a name, μελαμπυρον, used by Theophrastus)

melancholicus -a -um sad-looking, drooping, melancholy

melanciclus -a -um with dark circular markings

Melandrium Black-oak (the name used in Pliny)

Melanodiscus -a -um Black-disc (floral feature)

melanophloeus -a -um black-barked

melanops black-eyed

melanoxylon black-wooded

Melastoma Black-mouth (the fruits stain the lips black)

meleagris -is -e Greek name for Meleager of Calydon, chequered as is a guinea fowl (*Numidia meleagris*) and snake's-head fritillary (*Fritillaria meleagris*)

meles badger, *meles*

meli- honey-, μελι

Melia from the Greek name for ash (the resemblance of the leaves)

Melianthus Honey-flower

Melica Honey-grass

meliciferus -era -erum musical (*melicus*, musical)

Melilotus Honey-lotus (Theophrastus' name, μελιλωτος, refers to melilot's attractiveness to honeybees)

melinus -a -um quince-like, quince-coloured

Meliosma Honey-perfumed (the fragrance of the flowers)

Melissa Honeybee (named for the nymph who, in mythology, kept bees; and the plant's use in apiculture)

melissophyllus -a -um (mellisifolius) balm-leaved, with *Melissa*-like leaves

melitensis -is -e from Malta, Maltese

Melittis Bee (bastard balm attracts bees. A Greek derivation from *Melissa*)

melleus -a -um of honey (smelling or coloured)

mellifer -era -erum honey-bearing

mellinus -a -um the colour of honey

mellitus -a -um darling, honey-sweet

melo- melon-

Melocactus Melon-cactus (the shape)

melongena apple-bearer (producing a tree-fruit, the egg plant)

meloniformis -is -e (meloformis -is -e) like a ribbed-sphere, melon-shaped

Melothria the Greek name for bryony

membranaceus -a -um thin in texture, skin-like, membranous

Memecylon from the Greek name for the fruits of *Arbutus*, which are similar

memnonius -a -um dark brown, brownish-black, changeable

mene-, meni- moon-, crescent-

-mene membrane (μενινξ)

meniscatus -a -um curved-cylindrical

Menispermum Moon-seed, the compressed, curved stone of the fruit

Mentha the name in Pliny

menthoides mint-like, resembling *Mentha*

mentorensis -is -e from Mentor, Ohio, USA

Mentzelia for Christian Mentzel (1622–1701), early plant name lexicographer

Menyanthes Moon-flower (Theophrastus' name, μενανθος, for *Nymphoides*)

Menziesia for Archibald Menzies (1754–1842), English naturalist on the *Discovery*

meonanthus -a -um small-flowered
Mercurialis -is -e named by Cato for Mercury, messenger of the gods
Merendera from the Spanish vernacular name, quita meriendas
meri-, meros- partly-, part-
meridianus -a -um, meridionalis -is -e of noon, flowering at midday, southern
-meria, -meris -is -e -parts, -μερος
Mertensia for Franz Carl Mertens (1764–1831), German botanist
-merus -a -um -partite, -divided into, -merous, -μερος
merus pure, bare
mes-, mesi-, meso- middle-, μεσος, somewhat-
mesargyreus -a -um with silver towards the middle (leaf colouration)
Mesembryanthemum Midday-flower (flowers open in full sun)
messanius -a -um from Messina
-mestris -is -e -months (the period of growth or flowering)
Mezereum latinized from the Arabic, masarjun
mesoponticus -a -um from the middle sea (lakes of central Africa)
mesopotamicus -a -um from between the rivers
Mespilus Theophrastus' name, μεσπιλη σατανειος, for the medlar
messaniensis -is -e from Messina, Italy
messeniensis -is -e from Messenia, Morea, Greece
met-, meta- amongst-, next to-, after-, behind-, later-, with- (μετα)
metallicus -a -um lustrous, metallic in appearance
Metasequoia Close-to-*Sequoia* (resemblance of the dawn redwood)
meteoris -is -e dependent upon the weather (flowering)
methystico-, methysticus -a -um intoxicating
metro- mother-, μητηρ, centre-, heart-
Metrosideros Heart-of-iron (σιδηρος), the hard timber
Metroxylon Heart-wood (the large medulla)
Meum (Meon) an old Greek name, μηον, in Dioscorides
mexicanus -a -um from Mexico, Mexican
mezereum a name used by Avicenna (980–1037) from the Arabic (mazarjun)
Mibora an Adansonian name of uncertain meaning

micaceus -a -um from mica soils
micans shining, sparkling, glistening
Michauxia for André Michaux (1746–1803), French botanist
micr-, micra-, micro- small- (μικρο)
micranthus -a -um small-flowered
microbiota small-*Thuja* (*Biota* was an earlier synonym for *Thuja*)
Microcala Little-beauty
microcarpus -a -um small-fruited
microdasys small and hairy, with short shaggy hair
Microdesmis Small-clusters (refers to the clustered flowers)
microdon small-toothed
microglochin small-point (the tip of the flowering axis)
Microglossa Small-tongue (the short ligulate florets)
Microlepia Small-scale (thin outward-facing indusium is attached at the base and sides)
micromerus -a -um with small parts or divisions
Microsisymbrium Little-*Sisymbrium*
-mict- -mixed-, -mixture-
mikanioides resembling *Mikania* (climbing hemp-weed)
miliaceus -a -um millet-like, pertaining to millet
miliaris -is -e minutely glandular-spotted
militaris -is -e upright, resembling part of a uniform
Milium the Latin name for a millet grass
mille- a thousand- (usually means 'very many')
millefolius -a -um thousand-leaved (much divided leaves of milfoil)
Miltonia for Charles Fitzwilliam, Viscount Milton
mimetes mimicking
Mimosa Mimic (the sensitivity of the leaves, μιμος an imitator)
Mimulus Ape-flower (the flowers mimic a monkey's face)
miniatus -a -um cinnabar-red, the colour of red lead
minimus -a -um least, smallest
minor -or -us smaller
Minuartia for Juan Minuart (1693–1768), Spanish botanist
minus -a -um small

minutissimus -a -um extremely small, smallest
minutus -a -um very small, minute
mio- see *meio-*
Mirabilis, mirabilis -is -e Wonderful, extraordinary, astonishing
mirandus -a -um extraordinary
Miscanthus Stem-flower (μισχος), the elongate inflorescence
miser -era -erum wretched, inferior
Misopates Reluctant-to-open
missouriensis -is -e from Missouri, USA
Mitchellia for John Mitchell (d. 1772), botanist in Virginia, USA
Mitella Little-mitre (the shape of the fruit)
mithridatus -a -um for Mithridates Eupator, king of Pontus (mithridates give protection against poisons)
mitis -is -e gentle, mild, bland, not acid, without spines
Mitragyna Turban-shaped-ovary
Mitraria Capped (the bracteate inflorescence)
mitratus -a -um turbaned, mitred (μιτρα head-dress)
mitriformis -is -e, mitraeformis -is -e mitre-shaped, turban-shaped
mixo- mixing-, mingling- (μιξις)
mixtus -a -um mixed
mlokosewitschii for Herr Ludwig Mlokosewitsch, who found his *Paeonia* in the central Caucasus
-mnemon -fixed characters
mnio- moss-, *Mnium-*
Mniopsis Moss-like (genus of the aquatic *Podostemaceae*)
modestus -a -um modest, unpretentious
Moehringia for P.H.G. Möhring (1710–1792), German naturalist
Moenchia for Conrad Moench (1744–1805), German botanist
moesiacus -a -um from Moesia, Balkans
moldavicus -a -um from Moldavia, Danube area
molendinaceus -a -um, molendinaris -is -e shaped like a mill-sail
Molinia, molinae for Juan I. Molina (1740–1829), writer on Chilean plants
Molium Magic-garlic (after *Allium moly*)

molle from Peruvian name, mulli, for *Schinus molle*
mollearis -is -e resembling *Schinus molle*
molli-, mollis -is -e softly-hairy, soft
molliaris -is -e supple, graceful, pleasant
molliceps soft-headed
Mollugo Soft (a name in Pliny)
Moltkia for the Danish Count Joachim Gadske Moltke (d. 1818)
moluccanus -a -um from the Moluccas (Indonesia)
Moluccella derivation obscure (Bells of Ireland)
moly the Greek name of a magic herb
molybdeus -a -um, molybdos sad, neutral-grey, lead-coloured
mombin a West Indian vernacular name for hog plum, *Spondias mombin*
mon-, mona-, mono- one-, single-, alone- (μονος)
monadelphus -a -um in one group or bundle (stamens)
monandrus -a -um one-stamened, with a single stamen
Monarda for Nicholas Monardes of Seville (1493–1588), first herbal writer to include newly discovered American plants
mondo from a Japanese vernacular name
monensis -is -e from Anglesey or the Isle of Man, both formerly known as Mona
Moneses One-product (the solitary flower)
mongholicus -a -um, mongolicus -a -um from Mongolia, Mongolian
moniliformis -is -e necklace-like, like a string of beads
mono- single-, μονος-
monoclinus -a -um hermaphrodite, with stamens and ovary in one flower
monoclonos single-branched (–κλων)
monococcus -a -um one-fruited or -berried (–κοκκος)
Monodora Single-gift (the solitary flowers)
monoicus -a -um separate staminate and pistillate flowers on the same plant, moneocious
monorchis -is -e one-testicle (*Herminium* has a single tuber at anthesis)

Monotes Solitary (the first, and only genus of Dipterocarps in Africa when erected)
Monotropa One-turn (the band at the top of the stem)
monspeliensis -is -e, monspessulanus -a -um from Montpellier, S France
Monstera derivation uncertain, of huge size or monstrous foliage?
monstrosus -a -um, monstrus -a -um abnormal, monstrous, wonderful, horrible
montanus -a -um, monticolus -a -um of mountains, mountain-dweller
Montbretia for Antoine François Ernest Conquebert de Montbret (1781–1801), died in Cairo on the French expedition to Egypt
montevidensis -is -e from Montevideo, Uruguay
Montia for G.L. Monti (1712–1797), Italian botanist
monticolus -a -um mountain-loving
Moricandia for M. Etienne (Stephan) Moricand (1779–1854), Swiss botanist
morifolius -a -um mulberry-leaved, with *Morus*-like leaves
Moringa from an Indian vernacular name
morio madness
-morius -a -um -divisions, -parts, -merous (of the flower)
-morphus -a -um -shaped, -formed (μορφη)
morsus-ranae mouth of the frog (frog-bit)
mortefontanensis -is -e from the Chantrier brothers' nursery, Motrefontaine, France
mortuiflumis -is -e of dead water, growing in stagnant water
Morus the ancient Latin name for the mulberry
moschatellina an old generic name for *Adoxa* (musk-fragrant)
moschatus -a -um musk-like, musky (scented) (μοσκη)
moscheutos a vernacular name for swamp rose-mallow, *Hibiscus moscheutos*
mosaicus -a -um parti-coloured, coloured like a mosaic
moupinensis -is -e from Mupin, W China
moxa a vernacular name for the woolly leaves of *Artemisia moxa*
mucosus -a -um slimy

mucro-, *mucroni-* pointed-, sharp-pointed-

mucronatus -a -um with a hard sharp-pointed tip, mucronate (see Fig. 7(*b*))

mucronulatus -a -um with a hard, very short, pointed tip

Muehlenbeckia for Dr H. Gustave Muehlenbeck (1798–1845), Swiss physician

Muehlenbergia for Henri Ludwig Muehlenberg (1756–1817)

muelleri for Otto Ferdinand Mueller (1730–1784) or Ferdinand von Mueller (1825–1896)

mughus, *mugo* an old Italian vernacular name for the dwarf pine, *Pinus mugo*

Mulgedium Milker (Cassini's name refers to the possession of latex as in *Lactuca*)

mult-, *multi-*, *multus -a -um* many

multicavus -a -um with many hollows, many-cavitied

multiceps many-headed

multifidus -a -um much divided, deeply incised

multijugus -a -um pinnate, with many pairs of leaflets

multiplex with very many parts, very-double (flowered)

multiramosus -a -um many-branched

mume from the Japanese name, ume

mundulus -a -um quite neat, neatish

mundus -a -um clean, neat, elegant, handsome

munitus -a -um fortified, armed

muralis -is -e growing on walls, of the walls

muralius -a -um covering walls

muricatus -a -um rough with short superficial tubercles, muricate

murice from a vernacular name for the bark of *Byrsophyllum* species

murinus -a -um mouse-grey, of mice

murorum of walls

murra myrrh

Musa for Antonio Musa (63–14 BC), physician to Emperor Augustus

musaicus -a -um mottled like a mosaic, resembling *Musa*

musalae from Mt Musala, Bulgaria
Muscari Musk-like (from the Turkish, moscos – fragrance)
muscari- fly-, like *Muscari* inflorescence-
musci- fly-, moss-
muscifer -era -erum fly-bearing (floral resemblance)
musciformis -is -e moss-like
muscipulus -a -um fly-catching (*Dionaea muscipula*, Venus' flytrap)
muscivorus -a -um fly-eating
muscoides fly-like
muscosus -a -um moss-like, mossy
musi- banana-, *Musa-*
Mussaenda from a Sinhalese vernacular name
mussini for Count Grafen Apollos Apollosowitsch Mussin-Puschkin (d. 1805), phytochemist from the Caucasus (*Nepeta mussini*)
mutabilis -is -e changeable (in colour), mutable
mutans changing, variable, mutant
mutatus -a -um changed, altered
muticus -a -um without a point, not pointed, blunt
mutilatus -a -um roughly divided, as though torn
Mutisia (Mutisa) for Joseph Celestino B. Mutis y Bosio (1732–1808), Spanish discoverer of *Cinchona*
myagroides resembling *Myagrum*
Myagrum Mouse-trap (Dioscorides' name, μυαγρον)
Mycelis de l'Obel's name has no clear meaning
-myces, myco- -fungi, fungus-, mushroom-
myiagrus -a -um fly-catching (sticky)
myo- mouse-, closed-, (also muscle-, as in myocardial)
myoctonus -a -um mouse-death, poisonous to mice
myosorensis -is -e from Mysore, India
Myosotidium *Myosotis*-like
Myosotis Mouse-ear (Dioscorides' name, μυοσωτις)
Myosoton Mouse-ear (Dioscorides' name synonymous with *Myosotis*)
Myosurus Mouse-tail (the fruiting receptacle)

myr-, myro- myrrh-, *Myrrhis-*
myrianthus -a -um with a large number of flowers
Myrica Fragrance (the ancient Greek name, μυρικη, for *Tamarix*)
Myricaria *Myrica*-like, a Homeric name for a tamarisk
myrio- numerous-, myriad-
Myriophyllum Numerous-leaves (Dioscorides' name μυριοφυλλον)
Myristica Myrrh-fragrant (true nutmeg, *M. fragrans*)
myristicus -a -um myrrh-like (calabash-nutmeg, *Monodora myristica*)
myrmeco- ant- (many ant/plant symbioses involve plant structure modifications)
myrmecophilus -a -um ant-loving (plants with special ant accommodations and associations)
myrosmus -a -um myrrh-fragrant
Myrrhis Dioscorides' ancient name, μυρρηα, for true myrrh, *Myrrhis odorata*
myrrhus -a -um myrrh (*Commiphora myrrha*)
Myrsine Dioscorides' ancient name for the myrtle
myrsinites myrtle-like (μυρσινη-)
myrsinoides *Myrsine*-like
myrti- myrtle-, *Myrtus-*
myrtifolius -a -um myrtle-leaved
Myrtus the Greek name, μυρτον, for myrtle
mystacinus -a -um moustached, whiskered
mysurensis -is -e from Mysore, India
myurus -a -um, myuros mouse-tailed (μυς–ουρα)
Myurus Mouse-tail (the fruiting receptacle)
myx-, myxo- amoeboid-, mucus-, slime-

nacreus -a -um mother-of-pearl-like
naevosus -a -um freckled, with mole-like blotches
Naias, Najas one of the three mythological freshwater nymphs, or Naiads (see *Nymphaea* and *Nyssa*)
nairobensis -is -e from Nairobi, Kenya
nama-, namato- brook-

namaquensis -is -e from Namaqualand, western South Africa
namatophilus -a -um brook-loving
nan, nana-, nanae-, nani-, nano-, nanoe- dwarf
Nandina from its Japanese name, nandin
nanellus -a -um very dwarf
nannophyllus -a -um small-leaved
nanus -a -um dwarf
napaeifolius -a -um (napeaefolius -a -um) mallow-leaved, *Napaea*-leaved
napaulensis -is -e from Nepal, Nepalese
napellus -a -um swollen, turnip-rooted, like a small turnip
napi- turnip-
napifolius -a -um turnip-leaved
Napoleona (Napoleonaea) for Emperor Napoleon Bonaparte
Napus the name in Pliny for a turnip
narbonensis -is -e from Narbonne, S France
Narcissus the name of a youth in Greek mythology who fell in love with his own reflection, torpid (the narcotic effect)
Nardurus *Nardus*-tail (the narrow inflorescence)
Nardus Spikenard-like (the lower parts of *N. stricta* are a little like the biblical spikenard *Nardostachys jatamansi*)
narinosus -a -um broad-nosed
Narthecium Little-rod (the stem, also an anagram of *Anthericum*)
Nasturtium Nose-twist (from Pliny's *quod nasum torqueat*, the mustard-oil smell)
nasutus -a -um acute, large-nosed
nathaliae for Queen Natholia, wife of a former King of Milan
natans floating on water, swimming
Naumbergia for S.J. Naumberg (1768–1799), German botanist
nauseosus -a -um nauseating
nauticus -a -um, nautiformis -is -e boat-shaped
navicularis -is -e boat-shaped
nebrodensis -is -e from Mt Nebrodi, Sicily
nebulosus -a -um cloud-like, clouded, vaporous, nebulous
Nectaroscordum Nectar-garlic

neglectus -a -um (formerly) overlooked, disregarded, neglected
negundo from a Sanskrit name for a tree with leaves like box-elder
Neillia for Patrick Neill (1776–1851), Edinburgh botanist
Nelumbo from a Sinhalese name
-nema, nema-, nemato- -thread, thread-, thread-like-
Nemesia a name, νεμεσιον, used by Dioscorides for another plant
nemo- of glades-, glade- (νεμος, κνημος)
Nemopanthus (Nemopanthes) Thread-flower (the slender pedicels)
Nemophila Glade-loving (νεμος), woodland habitat
nemoralis -is -e, nemorosus -a -um, nemorum of woods, sylvan
nemossus -a -um from Clermont
nemusculus medieval Latin for underwood or scrub
neo- new-
neomontanus -a -um from Neuberg, Germany
neopolitanus -a -um from Naples, Neapolitan
Neoregelia for Eduard Albert von Regel
Neotinnea (Neotinea) New-*Tinnea* (for similarity to the genus named for three Dutch ladies who explored on the Nile)
Neottia Nest-of-fledglings (the appearance of the roots of *Neottia nidus-avis*, or 'bird's nest bird's nest')
nepalensis -is -e from Nepal, Nepalese
Nepenthes Euphoria (its reputed drug property of removing anxiety)
Nepeta the Latin name, from Nepi, Italy
nephr-, nephro- kidney-shaped-, kidney- (νεφρος)
Nephrodium Kidneys (the shape of the indusia of the sori)
nephroideus -a -um reniform, kidney-shaped
Nephrolepis Kidney-scale (the shape of the indusia of the sori)
nericus -a -um from the province of Närke, Sweden
nerii- oleander-like-, *Nerium-*
neriifolius -a -um (nereifolius -a -um) *Nerium*-leaved
Nerine a sea Nymph, daughter of Nereus
Nerium the ancient Greek name for oleander, *Nerium oleander*
Nertera Lowly (νερτερος), small stature
nerterioides resembling *Nertera* (bead plants)

nervalis -is -e loculicidal on the mid-rib, with a tendril-like prolongation of the mid-nerve

nervatus -a -um, *nervis -is -e* nerved or veined

nervosus -a -um with prominent nerves or veins

Neslia for the French botanist, Nesles

neso- island-

Nesogordonia Island-*Gordonia* (it was originally thought to be confined to Madagascar)

nesophilus -a -um island-loving

nessensis -is -e from Loch Ness, Scotland

-neurus -a -um -nerved, -veined

nevadensis -is -e from Nevada or the Sierra Nevada, USA

nicaensis -is -e from Nice, SE France or Nicaea, Bithynia, NW Turkey

Nicandra for Nicander of Calophon (100 BC), writer on plants

Nicotiana for Jean Nicot, who introduced tobacco to France in the late 16th century

nictitans moving, blinking, nodding

nidi-, *nidus* nest, nest-like

Nidularium Little-nest (the appearance of the compound inflorescence)

nidus-aves bird's-nest (resemblance)

Nierembergia for Juan Eusebia Nieremberg (1594–1658), Spanish Jesuit naturalist

Nigella Blackish (the seed coats)

nigellastrum medieval Latin name for corn-cockle

niger -ra -rum black

nigericus -a -um from Nigeria, West Africa

nigrescens darkening, turning black

nigri-, *nigro-* black-, dark-

nigricans almost black, blackish with age

nikoensis -is -e from Nike, Japan

niliacus -a -um from the River Nile

niloticus -a -um from the Nile Valley

nimus -a -um wooden

nipho- snow-

nipponicus -a -um from Japan (Nippon), Japanese

nissanus -a -um from Nish, SE Serbia

nissolia for Guillaume Nissole, 17th century botanist of Montpellier, France

nitens, *nitidi-*, *nitidus -a -um* glossy, with a polished surface, neat, shining

Nitraria Soda-producer (grows in saline deserts, when burnt yields nitre)

nitrophilus -a -um alkali-loving (growing on soda- or potash-rich soils)

nivalis -is -e growing near snow (*nix*, *nivis* snow)

niveus -a -um, *nivosus -a -um* purest white, snow-white

nobilis -is -e famous, grand, noble, notable

nocti- night-

noctiflorus -a -um, *nocturnus -a -um* night-flowering

nocturnalis -is -e at night, for one night (flowering)

nodiflorus -a -um flowering at the nodes

nodosus -a -um many-jointed, conspicuously jointed, knotty

nodulosus -a -um with swellings (on the roots), noduled

noeanus -a -um for Wilhelm Noe

Nolana Small-bell (*nola*)

noli-tangeri touch-not (the ripe fruit ruptures, expelling seed on touch)

noma-, *nomo-* meadow-, pasture-, νομη

Nomalxochia the Mexican vernacular name

nominius -a -um customary (νομιμος)

Nomocharis Meadow-grace (νομος)

non- not-, un-

nonpictus -a -um of plain colour, not painted

nonscriptus -a -um (nondescriptus -a -um) unmarked, not written upon

nootkatensis -is -e, *nutkatensis -is -e* from Nootka Sound, British Columbia, Canada

nordmannianus -a -um for Alexander von Nordmann (1843–1866), zoologist of Odessa and Helsingfors
normalis -is -e representative of the genus, usual, normal
norvegicus -a -um from Norway, Norwegian
notatus -a -um spotted, lined, marked
notho-, nothos-, nothus -a -um false-, spurious-, not-true-, bastard- (νοθος)
Nothofagus False-beech
Nothoscordum Bastard-garlic
noti-, notio- southern-, νοτος
noto- surface-, the back- (νωτον)
nov-, novae-, novi- new-
novae-angliae from New England
novae-belgii (novi-belgae) from New Belgium (New Netherlands or New York)
novae-caesareae (novi-caesareae) from New Jersey, USA
novae-zelandiae from New Zealand
noveboracensis -is -e from New York, USA
novem- nine-
novi-caesareae from New Jersey, USA
nubicolus -a -um, of cloudy places
nubicus -a -um from the Sudan (Nubia), NE Africa
nubigenus -a -um (nubiginus -a -um) cloud-formed, cloud-born
nubilorum from high peaks, of clouds
nubilus -a -um gloomy, sad, dusky, greyish-blue
nucifer -era -erum nut-bearing
nuculosus -a -um containing hard, nut-like seeds
nudatus -a -um, nudi-, nudus -a -um bare, naked
nudicaulis -is -e naked-stemmed, leafless
numidicus -a -um from Algeria (Numidia), Algerian
nummularis -is -e circular, coin-like (the leaves)
nummularius -a -um money-wort-like, resembling *Nummularia*
Nuphar the Persian name for a water-lily (ancient Latin *nenuphar, ninufer*)

nutabilis -is -e sad-looking, drooping, nodding
nutans drooping, nodding (the flowers)
nutkanus -a -um see *nootkatensis*
nux- nut-
nux-vomica with nuts causing vomiting (*Strychnos nux-vomica* contain the alkaloid strychnine)
-nychius -a -um -clawed
nyct-, nycto- night-
nyctagineus -a -um night-flowering
Nyctanthus Night-flower
nyctanthus -a -um nocturnal-flowering
Nyctocalos Night-beauty
nycticalus -a -um beautiful at night
Nymphaea for Nymphe, one of the mythological fresh-water Naiads
nymphae- waterlily-like-, *Nymphaea*-like-
Nymphe the name used by Theophrastus
Nymphoides Resembling-*Nymphaea*
Nyssa for Nyssa, one of the mythological fresh-water Naiads

ob-, oc-, of-, op- contrary-, opposite-, inverted-, inversely-, against-, completely-
obconicus -a -um like an inverted cone
obcordatus -a -um inversely cordate (stalked at narrowed end of a heart-shaped leaf), obcordate
obesus -a -um succulent, fat
obfuscatus -a -um clouded over, confused
Obione Daughter-of-the-Obi (a Siberian river)
oblanceolatus -a -um narrow and tapering towards the base
oblatus -a -um somewhat flattened at the ends, oval, oblate
obliquus -a -um slanting, unequal-sided, oblique
oblongatus-a -um, oblongi-, oblongus -a -um elliptic with blunt ends, oblong-
obovalis -is -e, obovatus -a -um egg-shaped in outline with the narrow end lowermost, obovate

obscissus -a -um with a squared-off end, cut off

obscurus -a -um dark, dingy, obscure, of uncertain affinity

obsoletus -a -um rudimentary, decayed

obstructus -a -um with the throat of the corolla restricted by hairs or appendages

obtectus -a -um covered over

obturbinatus -a -um reverse top-shaped, wide at the base and tapered to the apex

obtusatus -a -um, obtusi-, obtusus -a -um blunt, rounded, obtuse

obtusior more obtuse (than the type)

obvallaris -is -e, obvallatus -a -um (obvalearis) walled around, enclosed, fortified

obvolutus -a -um half-amplexicaule, with one leaf margin overlapping that of its neighbour

occidentalis -is -e western, occidental, of the West

occultus -a -um hidden

oceanicus -a -um growing near the sea

ocellatus -a -um (ocelatus -a -um) like a small eye, with a colour-spot bordered with another colour

Ochna an ancient Greek name, οχνη, used by Homer for a wild pear

ochnaceus -a -um resembling *Ochna*

ochr-, ochro- ochre-, pale-yellow- (ωχρος)

ochraceus -a -um ochre-coloured, yellowish

ochroleucus -a -um buff-coloured, yellowish-white (ωχρος–λευκος)

ochth-, ochtho- slope-, dyke-, bank-, οχθη

Ochthocosmus Hill-decoration (distinctive leaves, persistent flowers and montane habitat)

ocimoides, ocymoides sweet basil-like, resembling *Ocimum*

Ocimum the Greek name, οκιμον, for an aromatic plant

oct-, octa-, octo- eight-

octandrus -a -um eight-stamened

Octadesmia Eight-bundles (there are eight pollinial masses)

Octolepis Eight-scales (the paired scale-like petals)

Octolobus Eight-lobed (the calyx)

oculatus -a -um eyed, with an eye

oculus-christi Eye of Christ (*Inula oculus-christi*)

oculus-solis sun's-eye-

-odes -like, -resembling, -shaped, -similar to -οιδα

odessanus -a -um from Odessa, Black Sea area of Ukraine

odont-, odonto- tooth-, οδοντος

Odontites For-teeth (the name, οδους–ιτης, in Pliny refers to its use for treating toothache)

odontochilus -a -um with a toothed lip

Odontoglossum Toothed-tongue (the toothed lip)

odontoides tooth-like, dentate

odoratus -a -um, odorifer -era -erum, -odorus -a -um fragrant, -scented

-odus -a -um -joined

oedo- swelling-, becoming swollen- (οιδειν-)

Oedogonium Swollen-ovary (the enlarged gynoecial cells)

oelandicus -a -um from Öland, Sweden

Oenanthe Wine-fragrant

Oenothera Ass-catcher (the Greek name, οινοθηρας, for another plant but the etymology is uncertain)

officinalis -is -e, officinarum of the apothecaries, sold in shops, officinal medicines

Oftia a name by Adanson with no clear meaning

-oides, -oideus -a -um -like, -resembling, -shaped, οιδα

oistophyllus -a -um arrow-shaped-leaved

Olax Furrow (the appearance given by the two-ranked leaves)

olbia, olbios rich or from Hyères (*Olbia*), France

Olea Oily-one (the ancient name for the olive)

oleagineus -a -um, oleaginosus -a -um fleshy, rich in oil

oleander old generic name, ολεανδρη, used by Dioscorides (Italian, oleandra, for the olive-like foliage)

Olearia Olive-like (similarity of the leaves of some species)

Oleaster Olive-like (*Eleagnus*, Theophrastus used the name for a willow)

olei- olive-, *Olea*-

oleifer -era -erum oil-bearing
olens fragrant, musty, smelling
-olentus -a -um -fullness of, -abundance
oleospermus -a -um oil-seeded
oleraceus -a -um of cultivation, vegetable, aromatic, esculent
olgae for Olga Fedtschenko
olibanum from the Arabic, al luban, for the resinous secretion of *Boswellia*
olidus -a -um stinking, smelling
olig-, oligo- feeble-, few-
oliganthus -a -um with few flowers
oligospermus -a -um with few seeds
olisiponensis -is -e from Lisbon
olitorius -a -um of gardens or gardeners (*holitorius*), salad vegetable, culinary
olusatrum Pliny's name for a black-seeded pot-herb
olympicus -a -um from Mt Olympus, Greece, Olympian
ombro- rain-storm-, shade-
omeiensis -is -e from Mt Omei, Omei Shan, Szechwan, China
omeiocalamus *Calamus* of Mt Omei, Szechwan, China
omiophyllus -a -um lacking reduced (submerged) leaves
omorika from the Serbian name for *Picea omorika*
omphalo- navel- (ομφαλος)
Omphalodes Navel-like (ομφαλος) the fruit shape of navelwort
-on -clan, -family
onc-, onco- tumour-, hook- (ογκος)
Oncidium Tumour (ογκος), the warted crest of the lip
Oncoba from the Arabic name, onkob
onegensis -is -e from Onega, Russia
onites a name, ονος, used by Dioscorides (of an ass or donkey)
Onobrychis Ass-bray (a name, ονοβρυχις, in Pliny for a legume eaten greedily by asses)
Onoclea Closed-cup, ονος–κλειω (the sori are concealed by the rolled frond margins)

onomatologia the rules to be followed in forming names
Ononis the classical name, ονωνις, used by Dioscorides
Onopordum (on) Ass-fart, ονοπορδον (its effect on donkeys)
onopteris ass-fern, from a name used by Tabernamontana
Onosma Ass-smell (said to attract asses)
oo- egg-shaped-, ωον-
opacus -a -um dull, shady, not glossy or transparent
opalus from the old Latin name, opulus, for maple
operculatus -a -um lidded, with a lid
opertus -a -um hidden
ophio- snake-like, snake- (οφις)
Ophiobotrys Serpentine-raceme (the slender branches of the inflorescence)
ophiocarpus -a -um with an elongate fruit, snake-like-fruited
ophioglossifolius -a -um snake's-tongue-leaved
Ophioglossum Snake-tongue (appearance of fertile part of frond – adder's tongue fern)
Ophiopogon Snake-beard
Ophrys Eyebrow (the name, οφρυς, in Pliny)
-ophthalmus -a -um -eyed, -eye-like
opistho- back-, behind-
Oplismenus Weapon (οπλισμος)
opo- sap- (feeding, of parasites)
oporinus -a -um of late summer, autumnal (οπωρινος)
oppositi- opposite-, opposed-
-ops, opseo-, -opsis -is -e -eyed, -like, -looking like, -appearance of (οψις)
optimus -a -um the best
opuli- guelder rose-like
opulus an old generic name for the guelder rose
Opuntia Tournefort's name for succulent plants from Opous, Boeotea, Greece
opuntiiflorus -a -um (opuntiaeflorus -a -um) *Opuntia*-flowered
-opus -foot

orarius -a -um of the shoreline

orbicularis -is -e, orbiculatus -a -um disc-shaped, circular in outline, orbicular

orcadensis -is -e from the Orkney Isles, Orcadian

orchioides resembling *Orchis*

Orchis Testicle, ορχις (the shape of the root-tubers)

orculae- barrel-, cask-

oreadis -is -e of the sun, heliophytic (the Oreads were mythical mountain Nymphs)

oreganus -a -um, oregonensis -is -e, oregonus -a -um from Oregon, USA

orellana from a pre-Linnaean name for annatto, the red dye from *Bixa*

oreo-, ores-, ori mountain- (ορος,ορεος)

Oreodoxa Mountain-glory

oreophilus -a -um mountain-loving, montane

oresbius -a -um living on mountains

organensis -is -e from Organ Mt, New Mexico, USA or Brazil

orgyalis -is -e a fathom in length, about 6 feet tall (the distance from finger-tip to finger-tip with arms outstretched)

orientalis -is -e eastern, oriental, of the East

Origanum Joy-of-the-mountain (Theophrastus' name, οριγανον, for an aromatic herb)

-orius -a -um -able, -capable of, -functioning

Orixa from the Japanese name for *Orixa japonica*

ormo- necklace-like-, necklace-

ornatus -a -um adorned, showy

ornitho- bird-like-, bird-

Ornithogalum Bird-milk (Dioscorides' name, ορνιθογαλον, for a plant yielding bird-lime)

ornithopodioides, ornithopodus -a -um bird-footed, like a bird's foot (the arrangement of the fruits or inflorescence)

Ornithopus Bird-foot, ορνιθος–πους (the disposition of the fruits)

ornithorhynchus -a -um like a bird's beak

ornus from the ancient Latin for manna-ash, *Fraxinus ornus*

Orobanche Legume-strangler, οροβος–αγχω (one species parasitizes legumes – see also *rapum-genistae*)
orobus an old generic name, οροβος, for *Vicia ervilia*
orontium an old generic name, οροντιον, for a plant from the Orontes River, Syria
orophilus -a -um mountain-loving, montane (ορος)
orospendanus -a -um of mountains
orphanidium fatherless, unrelated
ortgeisii for Eduard Ortgeis (1829–1916), of Zurich Botanic Garden (*Oxalis ortgeisii* tree oxalis)
orth-, ortho- correct-, straight-, upright- (ορθος)
Orthila Straight (the style)
Orthocarpus Upright-fruit
orthocladus -a -um with straight branches
ortubae from the region of Lake Maggiore, Italy
orubicus -a -um from Oruba Island, Caribbean
orvala origin obscure, possibly from the Greek for a sage- (ορμιν) like plant
Oryza from the Arabic name, eruz
Oryzopsis *Oryza*-resembler
oscillatorius -a -um able to move about a central attachment, versatile
-osma -scented, fragrant- (οσμη))
osmo- thrust-, pressure- (ωσμος)
Osmanthus Fragrant-flower (for the perfumed *Osmanthus fragrans*)
Osmaronia Fragrant-*Aronia* (the derivation is doubtful)
Osmunda either for Osmund the waterman or for the Anglo-Saxon god of thunder, equivalent of the Norse, Thor
osseus -a -um of very hard texture, bony
ossifragus -a -um of broken bones (said to cause fractures in cattle when abundant in lime-free pastures)
osteo- bone-like-, bone- (οστεον)
Osteomeles Bone-apple (the hard fruit)
Osteospermum Bone-seed (the hard-coated fruits)
ostiolatus -a -um having a small opening or mouth

ostraco- hard-shelled-

Ostrowskia for Michael Nicholajewitsch von Ostrowsky, Minister of the Russian Imperial Domains and botanist

ostruthius -a -um purplish

Ostrya a name, οστρυς, in Pliny for a hornbeam

-osus -a -um -abundant, -large, -very much

-osyne, -otes -notably

ot-, oto- ear-like-, ear- (ους,ωτος)

Otanthus Ear-flower (the shape of the corolla)

-otes, -otus -a -um -looking-like, -resembling, -having

Othonia Cloth-napkin (the covering of downy hairs)

otites relating to ears, an old generic name, from Rupius

otrubae for Joseph Otruba (b. 1889), of Moravia

Ottelia from the native Malabar name

Oubanguia from the name of the River Oubangui, Nigeria

ouletrichus -a -um with curly hair

Ouratea from the South American native name

Ourisia for Governor Ouris of the Falkland Islands

ovali-, ovalis -is -e egg-shaped (in outline), oval

ovati-, ovatus -a -um egg-shaped (in the solid or in outline), with the broad end lowermost

ovifer -era -erum, oviger -era -erum bearing eggs (or egg-like structures)

oviformis -is -e egg-shaped (in the solid), ovoid

ovinus -a -um of sheep

Oxalis Acid-salt (the name, οξαλις, in Nicander refers to the taste of sorrel)

oxodus -a -um of acid humic soils, οξωδης

oxy-, -oxys acid-, sharp-, -pointed (οξυς)

Oxyacantha Sharp-thorn (Theophrastus' name)

oxyacanthus -a -um having sharp thorns or prickles

oxycarpus -a -um having a sharp-pointed fruit

Oxycedrus Pungent-juniper

Oxycoccus Acid-berry

Oxydendrum (on) Sour-tree (the acid taste of the leaves), sourwood
oxygonus -a -um with sharp angles, sharp-angled
oxylobus -a -um with sharp-pointed lobes
oxylophilus -a -um of humus-rich soils, humus-loving
Oxypetalum Sharp-petalled
oxyphilus -a -um of acidic soils, acid soil-loving
Oxyria Acidic (the taste)
Oxytropis Sharp-keel (the pointed keel petal)
Ozothamnus Fragrant-shrub

pabularis -is -e, pabularius -a -um of forage or pastures
pachy- stout-, thick- (παχυς)
pachyphloeus -a -um thick-barked
Pachyphragma Stout-partition (the ribbed septum of the fruit)
Pachypodanthium Thick-footed-flowers (the crowded stalkless carpels)
Pachysandra Thick-stamens (the filaments)
(Pachystema, Pachistima, Pachystigma) see *Paxistima*
pacificus -a -um of the western American seaboard
padi- *Prunus padus-*
padus Theophrastus' name, παδος, for St Lucie cherry or from the River Po, Italy
Paederia Bad-smell (the crushed flowers)
Paeonia named by Theophrastus for Paeon, the physician to the Gods who, in mythology, was changed into a flower by Pluto
paganus -a -um of country areas, from the wild
pago- foothill-, παγος
palaestinus -a -um from Palestine, Palestinian
paleaceus -a -um covered with chaffy scales, chaffy
palaeo- ancient-
palinuri from Palinuro, Italy
Palisota for A.M.F. Palisot de Beauvois (1752–1820), French botanist
Paliurus the ancient Greek name for Christ-thorn

pallasii for Peter Simon Pallas (1741–1811), German naturalist and explorer

pallens pale

pallescens (palescens) becoming pale, fading

palliatus -a -um cloaked, hooded

pallidus -a -um greenish, somewhat pale

palmaris -is -e of a hand's breadth, about 3 inches wide

palmati-, *palmatus -a -um* with five or more veins arising from one point (usually on divided leaves), hand-shaped, palmate (see Fig. 5(*a*))

palmensis -is -e from Las Palmas, Canary Isles

palmi- date-palm-, palm-of-the-hand-

palmitifidus -a -um palmately incised

palpebrae eyelashed, with fringe of hairs

paludis -is -e of swamps

paludosus -a -um growing in boggy or marshy ground

palumbinus -a -um lead-coloured (the colour of wood-pigeons)

paluster -tris -tre of swampy ground (*palustris* is often used as a masculine ending in botanical names)

pamiricus -a -um of the Pamir Mountains, Tadzhikstan

pampini- tendrillar-, tendril-

pampinosus -a -um leafy, with many tendrils

pan-, *panto-* all- (παν, παντος)

panaci- *Panax-*

panamensis-is-e from Panama, Central America

Panax Healer-of-all (the ancient virtues of ginseng)

pancicii for Joseph Panĉić (1814–1888), Yugoslavian botanist

Pancratium All-potent (a name, πανκρατος, used by Dioscorides)

pandani- *Pandanus-*

Pandanus Malayan name, pandan, for screw-pines

pandorana Pandora's (objects of desire, the changing form of *Pandorea pandorana*)

Pandorea Pandora (Wonga wonga vine)

panduratus -a -um fiddle-shaped, pandurate, panduriform

paniceus -a -um like millet grain
paniculatus -a -um with a branched-racemose or cymose inflorescence, tufted, paniculate (see Fig. 2(*c*))
Panicum the ancient Latin name for the grass *Setaria*
panneformis -is -e with a felted surface texture, *pannus*, cloth
pannifolius -a -um cloth-leaved
pannonicus -a -um from Pannonia, SW Hungary
pannosus -a -um woolly, tattered, coarse, ragged
panormitanus -a -um from Palermo, πανορμος, Sicily
pantothrix hairy all round
Papaver the Latin name for poppies, including the opium poppy
papaya from a vernacular name for pawpaw, *Carica papaya*
paphio- Venus'-
Paphiopedilum Venus'-slipper (see *Cypripedium*), Venus' temple was at Paphos
papil-, papilio- butterfly-
papillifer -era -erum, papilliger -era -erum producing or bearing papillae
papillosus -a -um covered with papillae or minute lobes, papillate
pappi-, pappus- downy-, down-
papposus -a -um downy
papuli- pimple-
papulosus -a -um pimpled with small soft tubercles
papyraceus -a -um with the texture of paper, papery
papyrifer -era -erum paper-bearing
Papyrus Paper (the Greek name for the paper made from the Egyptian bulrush, *Cyperus papyrus*)
para- near-, beside-, wrong, irregular- (παρα)
parabolicus -a -um ovate-elliptic, parabolic in outline
paradisi, paradisiacus -a -um of parks, of gardens, of paradise
Paradisea for Count Giovani Paradisi
paradoxus -a -um strange, unusual, unexpected (παρα–δοξος)
paraguariensis -is -e, paraguayensis -is -e from Paraguay
paralias seaside, by the beach (ancient Greek name for a plant, παραλιος)

Parapholis Irregular-scales (the position of the glumes)
parasiticus -a -um living at another's expense, parasitic (formerly applied to epiphytes)
parellinus -a -um, parellus -a -um litmus-violet (lichen *Lecanora parella*)
parci- with few-
parcifrondiferus -a -um bearing few or small leafy shoots, with few-leaved fronds
pardalianches, pardalianthes leopard-strangling (a name in Aristotle, παρδαλιαγχες, for plants poisonous to wild animals. Leopard's-bane)
pardalinus -a -um, pardinus -a -um spotted or marked like a leopard (παρδαλις)
pardanthinus -a -um resembling *Belamcanda* (*Pardanthus*)
Parentucellia for Th. Parentucelli (Pope Nicholas V)
pari- equal-, paired-
parietalis -is -e, parietarius -a -um, parietinus -a -um of walls, parietal (also, the placentas on the wall within the ovary)
Parietaria Wall-dweller (a name in Pliny used for a plant growing on walls)
Parinari from a Brazilian vernacular name
paripinnatus -a -um with an equal number of leaflets and no odd terminal one
Paris Equal (the regularity of its leaves and floral parts)
parisiensis -is -e French (continental)
Parkinsonia for John Parkinson (1569–1629), author of *Paradisi in Sole*
parmularius -a -um like a small round shield
parmulatus -a -um with a small round shield
parnassi, parnassiacus -a -um from Mt Parnassus, Greece
Parnassia l'Obel's name for *Gramen Parnassi* – grass of Parnassus
Parochetus Brookside (οχετυς)
Parodia for Dr L.R. Parodi of Buenos Aires, writer on grasses
Paronychia Beside-nail (formerly used to treat whitlows)
Paropsis Dish-of-food

parqui from the Chilean name for *Cestrum parqui*
Parrotia for F.W. Parrot (1792–1841), German naturalist and traveller
parthenium an old generic name, παρθενιον, for composites with white ray florets. Virginal
Parthenocissus Virgin-ivy (French name Virginia creeper)
parthenus -a -um virgin, of the virgin, virginal (παρθενος)
-partitus -a -um -deeply divided, -partite, -parted
-parus -a -um -bearing, -producing
parvi-, parvus -a -um small-
parvulus -a -um very small, least
pascuus -a -um of pastures
Paspalum a Greek name for millet grass
Passiflora Passion-flower (the signature of the numbers of parts in the flower related to the events of the Passion)
Pastinaca Food, eatable, from a trench in the ground (formerly for carrot and parsnip)
pastoralis -is -e, pastoris -is -e growing in pastures, of shepherds
patagonicus -a -um from Patagonian area of South America
patavinus -a -um from Padua, Italy
patellaris -is -e, patelliformis -is -e knee-cap-shaped, small dish-shaped
patens, patenti- spreading out from the stem, patent
pateri- saucer-
patientia patience (corruption of patience dock *Lapathum*)
patulus -a -um spreading, opened up
pauci-, paucus -a -um little-, few
pauciflorus -a -um few-flowered
Paulownia for Princess Anna Pavlovna (Paulowna) (1795–1865), of The Netherlands, daughter of Czar Paul I of Russia
pauper-, pauperi- poor-
pauperculus -a -um of poor appearance
pausiacus -a -um olive-green
Pavonia, pavonianus -a -um for Don José Pavón (1790–1844), Spanish botanist in Peru

pavonicus -a -um, *pavoninus -a -um* peacock-blue, showy
pavonius -a -um peacock-blue, resembling *Pavonia*
Paxistima Short-stigma (παχις) the short style of the immersed ovary
pecten-veneris Venus' comb (a name used in Pliny)
pectinatus -a -um comb-like (scalloped), pectinate
pectinifer -era -erum with a finely divided crest, comb-bearing
pectoralis -is -e of the chest (used to treat coughs)
pedalis -is -e about a foot in length or stature
ped- stalk-, foot-
pedati- (pedali-), *pedatus -a -um* palmate but with the lower lateral lobes divided, pedate (see Fig. 5(*b*))
pedatifidus -a -um divided nearly to the base in a pedate manner
pedemontanus -a -um from Piedmont, N Italy (foot of the hills)
pedialis -is -e with a long flower-stalk
pedicellatus -a -um, *pedicellaris -is -e (pediculatus)* each flower clearly borne on its own individual stalk in the inflorescence, pedicellate
Pedicularis Louse-wort
pedicularis -is -e of lice (name of a plant in Columella thought to be associated with lice)
pedifidus -a -um shaped like a (bird's) foot
pedil-, *pedilo-* shoe-, slipper-, πεδιλον
Pedilanthus Shoe-flower (involucre of bird cactus)
pediophilus -a -um growing in upland areas
peduncularis -is -e, *pedunculatus -a -um* with the inflorescence supported on a distinct stalk, pedunculate
pedunculosus -a -um with many or conspicuous peduncles
Peganum Theophrastus' name for rue
pekinensis -is -e from Pekin (Beijing), China
pel- through-
Pelargonium Stork, πελαργος (Greek name compares the fruit shape of florists' Geranium with a stork's head)
pelegrina from a vernacular name for *Alstroemeria pelegrina*
pelicanos pelican-like

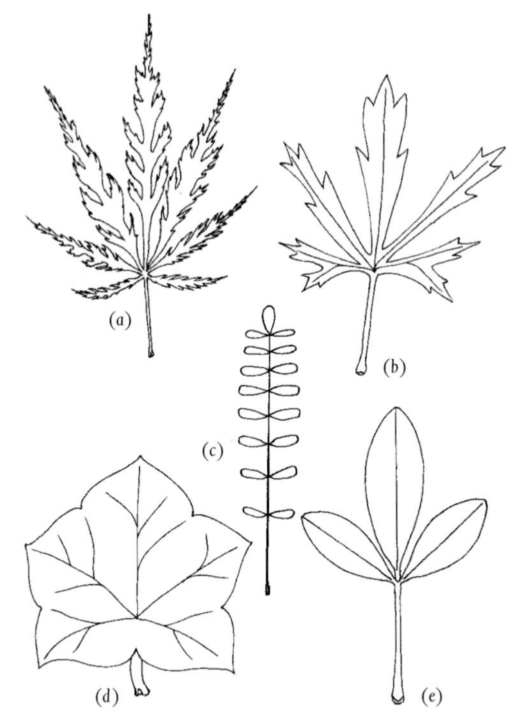

Fig. 5. Some leaf shapes which provide specific epithets.
(*a*) Palmate (e.g. *Acer palmatum* Thunb. '*Dissectum*'. As this maple's leaves mature, the secondary division of the leaf-lobes passes through incised-, *incisum*, to torn-, *laciniatum*, to dissected-, *dissectum*, -lobed, from one central point;
(*b*) pedate (e.g. *Callirhoe pedata* Gray). This is distinguished from palmate by having the lower, side lobes themselves divided;
(*c*) Pinnate (e.g. *Ornithopus pinnatus* Druce). When the lobes are more or less strictly paired it is called paripinnate, when there is an odd terminal leaflet it is called imparipinnate and when the lobing does not extend to the central leaf-stalk it is called pinnatifid;
(*d*) peltate (e.g. *Pelargonium peltatum* (L.) Ait.) has the leaf-stalk attached on the lower surface, not at the edge;
(*e*) Ternate (e.g. *Choisya ternata* H.B.K.). In other ternate leaves the three divisions may be further divided, ternately, palmately, or pinnately.

peliorrhincus -a -um like a stork's beak

pelios- black-, livid-

pelisserianus -a -um for Guillaume Pelisser, 16th century bishop of Montpellier, mentioned by Tournefort as discoverer of *Teucrium scordium* and *Linaria pelisseriana*

Pellaea Dusky (πελλος), the fronds of most

pellitus -a -um skinned, covered with a skin-like film

pellucidus -a -um through which light passes, transparent, clear, pellucid

pelochtho- mud-bank-

pelorius -a -um monstrous, peloric (e.g. radial forms of normally bilateral flowers)

pelta-, pelti-, pelto- shield-

peltafidus -a -um with peltate leaves that are cut into segments

peltatus -a -um stalked from the surface (not the edge), peltate (see Fig. 5(*d*))

Peltiphyllum Shield-leaf (the large leaves that follow the flowers). See *Darmera*

Peltophorum Shield-bearer (the shape of the stigma)

peltophorus -a -um with flat scales, shield-bearing

pelviformis -is -e shallowly cupped, shaped like a shallow bowl

pemakoensis -is -e from Pemako, Tibet

pen-, pent-, penta- five-

pendens, penduli-, pendulinus -a -um, pendulus -a -um drooping, hanging down

penduliflorus -a -um with pendulous flowers

penicillatus -a -um, penicillius -a -um (penicellatus) covered with tufts of hair, brush-like

peninsularis -is -e living on a peninsula

penna-, penni- feather-, feathered-, winged-

pennatus -a -um, penniger -era -erum arranged like the barbs of a feather, feathered

pennatifidus -a -um pinnately divided

penninervis -is -e pinnately nerved

Pennisetum Feathery-bristle
pennsylvanicus -a -um, pensylvanicus -a -um from Pennsylvania, USA
pensilis -is -e hanging down, pensile
Penstemon (Pentstemon) Five-stamens (five are present but the fifth is sterile)
penta- five- (πεντε)
pentadelphus -a -um with the stamens arranged in five bundles
Pentadesma Five-bundles (the grouping of the many stamens)
Pentaglottis Five-tongues (the scales in the throat of the corolla)
pentandrus -a -um with five stamens in the flower
pentapterus -a -um with five wings (e.g. on the fruit)
Pentas Five-fold
Penthorum Five-columns (the beaks on the fruit)
Peperomia Pepper-like (some resemble *Piper*)
peplis Dioscorides' name, πεπλις, for a Mediterranean coastal spurge
peploides *Peplus*-like, spurge-like
peplus Dioscorides' name, πεπλος, for a northern equivalent of *peplis*
Pepo Sun-cooked (Latin name for a pumpkin, ripening to become edible, πεπων)
per-, peri around-, through-, beyond-, extra-, very-, περι
peramoenus -a -um very beautiful, very pleasing
Peraphyllum Much-leaved (the crowded foliage)
percarneus -a -um deep-red
percurrens running through, along the whole length
percursus -a -um running through the soil
percussus -a -um actually or appearing to be perforated, striking
peregrinus -a -um strange, foreign, exotic
perennans, perennis -is -e through the years, continuing, perennial
perennitas continuing, of the perennial state
Pereskia for Nicholas Claude Fabry de Pieresc (1580–1637)
perfoliatus -a -um, perfossus -a -um the stem appearing to pass through the completely embracing leaves

perforatus -a -um pierced or apparently pierced with small round holes
perfossus -a -um pierced through, perfoliate
perfusus -a -um poured over, completely covered
pergamenus -a -um with a texture like that of parchment, of Pergamo, a town in Mysia
Pergularia Arbour (the twining growth)
peri- around-, about-
periclymenus -a -um Dioscorides' name, περικλυμενον, for a twining plant
Periploca Twine-around (the twining habit)
perlarius -a -um, perlatus -a -um with a pearly lustre, having pearl-like appendages
permiabilis -is -e penetrable
permixtus -a -um confusing
permutatus -a -um completely changed
Pernettya for A.J. Pernetty (1716–1801), accompanied Bougainville and wrote *A Voyage to the Faukland Islands*
peronatus -a -um booted, *pero*, with a woolly-mealy covering (on fungal fruiting bodies)
Perovskia (Perowskia) for P.A. Perovski (*c.* 1840), provincial governor of Orenburg, Russia
perpelis -is -e living on rocks which turn to clay
perpropinquus -a -um very closely related
perpusillus -a -um exceptionally small, very small, weak
perralderianus -a -um for Henri Réné le Tourneaux de la Perrauddière
Persea ancient Greek name for an oriental tree, Perseus hero of Greek legend
persi-, persici-, persicoides peach-
Persicaria Peach-like (the leaves)
persicarius -a -um resembling peach (the leaves), an old name for *Polygonum hydropiper*
persicus -a -um from Persia, Persian
persistens persistent

persolutus -a -um loose, lazy, free, rank

personatus -a -um with a two-lipped mouth, masked

perspicuus -a -um transparent, clear, bright

persutus -a -um perforate, with slits or holes

pertusus -a -um pierced through, perforated, dotted

Pertya for A.M. Perty (1800–1884), Professor of Natural History, Berne, Switzerland

perulatus -a -um wallet-like, with conspicuous scales (e.g. on buds)

perutilis -is -e always ready

peruvianus -a -um from Peru, Peruvian

perviridis -is -e deep-green

pes-, -pes -stalk, -foot

pes-caprae (pes-capriae) nanny-goat's foot (leaf shape of *Oxalis pes-caprae*)

pes-tigridis tiger's foot

-petalus -a -um -petalled

Petasites Wide-brimmed-hat (Dioscorides' name, πετασος, refers to the large leaves)

petaso- wide-brimmed, parachute-like-

petecticalis -is -e blemished with spots

petiolaris -is -e, petiolatus -a -um having a petiole, not sessile, distinctly petiolate

petiolosus -a -um with conspicuous petioles

petr-, petra-, petro- rock-like-, rock-, πετρος

petraeus -a -um rocky, of rocky places

petrodo- rock-strewn-area-

Petrophila Rock-lover (habitat preference)

Petrophytum Rock-plant (the habitat)

Petroselinum Dioscorides' name, πετροσελινον, for parsley

petroselinus -a -um parsley-like

Petunia from the Brazilian name, petun, for tobacco

Peucedanum a name, πευκεδανον, used by Theophrastus for hog fennel

Peyrousea see *Lapeirousia*

pezizoideus -a -um cup-shaped, orange-coloured (as the fungus *Peziza aurantia*)

Phacelia Bundle, φακελος

phaedr-, phaedro-, phaidro- gay-, φαιδρος

Phaedranthus Gay-flower (the colourful flowers of the climber *P. buccinatorius*)

phaen-, phaeno- shining-, apparent-, obvious- (φαινω, anglicized to *phan-*)

Phaenocoma Shining-hair (the large red flower-heads with spreading purple bracts)

phaeo-, phaeus -a -um dark-, dusky-brown, φαιος

Phaeomeria Dark-(purple)-parts

phaio-, Phaius dark-coloured (φαιος)

Phalaenopsis Moth-like (flowers of the moth orchid), φαλαινα

Phalaris Helmet-ridge (Dioscorides' name, φαλαρις, for a plume-like grass)

phaleratus -a -um (phalleratus) shining-white, ornamental, decorated (wearing medals), φαληρος

phalliferus -a -um bearing a crest, crested

phanero- conspicuous-, manifest-, visible-

Phaseolus Dioscorides' name for a kind of bean

Phegopteris Oak-fern (a name created by Linnaeus from φηγος, an oak)

Phelipaea (Phelypaea) for Louis Phelipeaux, Count Ponchartrain, patron of J.P. de Tournefort

Phellandrium (Phellandrion) a name in Pliny for an ivy-leaved plant

phello-, phellos corky-, cork (φελλος)

Phellodendron Cork-tree (the thick bark of the type species)

phen- see *phaen-*

phil-, philo-, -philus -a -um loving-, liking-, -fond of (φιλεω)

philadelphicus -a -um from Philadelphia

Philadelphus Brotherly-love (φιλαδελφος)

philaeus -a -um ground-hugging, earth-loving

Philesia Loved-one

philippensis -is -e from the Philippines
philonotis -is -e moisture-loving (φιλος–νοτις)
Phillyrea from an ancient Greek name
-philus -a -um -loving (φιλος, a friend)
phleb- vein- (φλεψ-, φλεβος)
phlebanthus -a -um with veined flowers
phleioides rush-like, resembling the grass *Phleum*
Phleum Copious (Greek name, φλεως, for a kind of dense-headed rush)
-phloebius -a -um -veined
-phloem with veined flowers
-phloeus -a -um -barked, -bark
phlogi- flame-, *Phlox*-like, φλοξ, φλογος
Phlomis Flame, φλομις (the hairy leaves were used as lamp wicks)
Phlox Flame (Theophrastus' name, φλοξ, for a plant with flame-coloured flowers)
phocaena seal or porpoise
phoeniceus -a -um scarlet, red with a little yellow (φοινιξ, φοινικος)
phoenicius -a -um from Tyre and Sidon (Phoenicia)
phoenicolasius -a -um red-purple-haired
Phoenix Phoenician (who introduced the date palm to the Greeks)
Pholiurus Scale-tail (φολις–ουρα)
Phoradendron Thief-tree (the parasitic habit)
Phormium Basket (the leaf-fibres were used for weaving), φορμιον
-phorus -a -um -bearing, -carrying (φορεο to carry)
Photinia Shining-one (φοτεινος), from the lustrous foliage
phoxinus minnow
phragma-, -phragma fence-, enclosure- (φραγμα)
Phragmites Hedge-dweller (common habitat)
phrygius -a -um from Phrygia, Asia Minor
phu foul-smelling
Phuopsis Valerian-like, resembling *Valeriana phu*
-phyceae, phyco- sea-weed-
Phygelius Fugitive, φυγας

Phyla derivation uncertain, φυλη tribe

Phylica Leafy (a name, φυλικη, in Theophrastus for a plant with copious foliage)

phylicifolius -a -um with leaves like those of *Phylica*

phyll-, phylla-, phyllo- leaf-, φυλλον

Phyllanthus Leaf-flower (some flower from edges of leaf-like phyllodes)

Phyllitis Dioscorides' name, φυλλιτις, refers to the simple leaf-like frond

phyllo- leaf- (φυλλον)

phyllobolus -a -um leaf-shedding, throwing off leaves

Phyllocladus (os) Leaf-branch (the flattened leaf-like cladodes)

Phyllodoce the name of a sea nymph

phyllomaniacus -a -um excessively leafy, a riot of foliage

Phyllostachys Leafy-spike (the leafy inflorescence)

-phyllus -a -um -leaved

Phymatodes Verrucosed (the sori are in depressions)

phymatodeus -a -um warted, verrucose

physa-, physo- bladder-, inflated-, bellows-, φυσα

Physalis Bellows (the inflated fruiting calyx resembles a bellows)

Physocarpus Bladder-fruit

physodes puffed out, inflated-looking

Physospermum Inflated-seed (fruit of bladder seed)

phyt-, -phyta, phyto- plant-, φυτον

Phyteuma a name, φυτευμα, used by Dioscorides

Phytolacca Plant-dye (the sap of the fruit)

pica magpie

Picea Pitch (the ancient Latin name, *pix*, refers to the resinous product)

piceus -a -um pitch-black, blackening

pichtus black (mis-spelling for *pictus*?)

Picrasma Bitterness (the bitter-tasting bark)

picridis -is -e ox-tongue-like, of *Picris*

Picris Bitter (Theophrastus', πικρις, name for a bitter potherb)

picro-, *-picron* bitter-, -bitter, πικρια, πικρος
picturatus -a -um variegated, picture-like
pictus -a -um brightly marked, painted, ornamental
Pieris from a name, πιεριδες, for the Muses of Greek mythology
pilaris -is -e pilose
Pilea Cap, *pileus*
pileatus -a -um capped, having a cap (*pileus*)
pileo- cap-
Pileostegia Cap-covered
piliferus, *pilifer -era -erum* bearing hairs (*pili*), with short soft hairs, ending in a long fine hair
pilo- felted with long soft hairs, πιλος
Pilosella Slightly soft-haired (Rufinus' name for *Hieracium pilosella*)
piloselloides hawkweed-like, *Pilosella*-like
pilosellus -a -um tomentose, finely felted with soft hairs
pilosiusculus -a -um hairy-ish, with sparse, very fine hairs
pilosus -a -um covered with soft distinct hairs, pilose
Pilularia Small-balls (*pilula*, a small ball), the shape of the sporocarps
pilularis -is -e, *pilulifer -era -erum* bearing small balls, glands or globular structures
Pimelea Fat (πιμελη), the oily seeds
Pimenta the Spanish name, pimento, for allspice, *Pimenta officinalis*
pimentoides allspice-like, *Pimenta*-like
Pimpinella a medieval name of uncertain meaning, from medieval French, pimprinele
pimpinellifolius -a -um *Pimpinella*-leaved
pinaster Pliny's name for *Pinus sylvestris*
pindicola, *pindicus -a -um* from the Pindus Mountains of N Greece
pinetorum of pine woods
pineus -a -um cone-producing, of pines, resembling a pine
pingui- fat-
Pinguicula Rich (*pinguis*), the fatty appearance of the leaves
pini- pine-like, pine-

pinnati-, *pinnatus -a -um* set in two opposite ranks, winged, feathered, pinnate (see Fig. 5(*c*))
pinnatifidus -a -um pinnately divided
pinsapo from the Spanish name, pinapares, for *Abies pinsapo*
Pinus the ancient Latin name for a pine
Piper from the Indian name for pepper
piperascens pepper-like, resembling *Piper*
piperatus -a -um, *piperitus -a -um* with a hot biting taste, peppered, pepper-like (the taste)
piperinus -a -um peppery (-scented)
Piptanthus Falling-flower (πιπτω to fall), quickly deciduous floral parts
Piptostigma Falling-stigma (the stigma falls off after flowering)
piri- pear-
piriformis -is -e pear-shaped
Pirola Small-pear (similarity of foliage)
Pirus the Latin name for a pear tree
piscinalis -is -e of ponds or pools, *piscena*
pisi-, *piso-* pea-like-, pea-
pisifer -era -erum bearing peas
pissardii (pissardi, pissarti) for M. Pissard who introduced *Prunus cerasifera* 'Pissardii'
Pistacia the Greek name, πιστακε, used by Nicander in 200 BC
Pistia Watery (habitat of the water lettuce) πιστος, derived from the Persian, foustag
Pisum the Latin name for the pea
pitanga a South American Indian name for *Eugenia pitanga*
pithece-, *pitheco-* ape-, monkey-
Pithecellobium Monkey-ears (the shape of the fruit)
Pittosporum Pitch-seed (the resinous coating of the seed)
pitui- mucus-, phlegm, *pituita*
pityophyllus -a -um with pine-like foliage
pityro- husk-, scurf-, πιτυρα

Pityrogramma Scurf-covered (lower surface of fronds becomes obscured by rod-like scaly secretions)

-pitys, pityoides pine-like

Pixidanthera Box-anthers (they dehisce with a lid, see *pyxidatus*)

placatus -a -um quiet, calm, gentle

placenti- placenta-, flat-cake-

placo- flat-, πλαξ

Placodiscus Flat-disc (the floral disc)

plagi-, plagio- oblique- (πλαγιος)

Plagiomnium Oblique-*Mnium*

Plagiostyles Oblique-styled (the short, fat stigma is to one side of the ovary)

-planatus -a -um -sided

Planera for J.J. Planer (1743–1789), Professor of Medicine at Erfurt, Germany

planeta, planetes not stationary, planet-like, wandering

plani- flat-, even-

planiusculus -a -um flattish, somewhat flat

plantagineus -a -um (plentigineus) ribwort-like, plantain-like

Plantago Foot-sole (ancient Latin, the way the leaves of some lie flat on the ground)

planus -a -um flat-, smooth

plasmo-, plasmodio- cytoplasm-

plat-, platy- broad-, wide-, flat- (πλατυς)

platanoides plane-tree-like, *Platanus*-like

Platanthera Flat-anthers

Platanus Flat-leaf, Flat-crown (the Greek name, πλατανος, for *Platanus orientalis*)

platy- broad- (πλατυς)

Platycarya Broad-nut (the compressed nutlet)

platycentrus -a -um wide-eyed, broad-centred

Platycerium Broad-horned (the stag's horn-like, dichotomous lobing of the fertile fronds)

Platycodon Wide-bell (the flower form)

Platycrater Wide-bowl (the broad calyx of the sterile flowers)
plebio-, *plebius -a -um* common, inferior
pleco- plaited- (πλεκω)
plecto-, *plectus -a -um* woven-, twisted- (πλεκτος)
Plectranthus Spurred-flower
plectro-, *plectrus -a -um* spur-, spurred
pleni-, *plenus -a -um* double, full
pleniflorus -a -um double-flowered
plenissimus -a -um very full or double-flowered
pleio-, *pleo-* many-, several-, full-, large-, thick-, more- (πλειος)
Pleione mother of the Pleiades in Greek mythology
pleiospermus -a -um thick-seeded
Pleiospilos Many-spotted (the leaves)
plesio- near to-, close by- (πλησιος)
pleura-, *pleuri-*, *pleuro-* ribs-, edge-, side-, of the veins- (πλευρα)
plexi-, *-plexus -a -um* knitted-, -braided, -network (πλεξι)
plicati-, *plicatus -a -um* folded-together-, -doubled, -folded (*plico* to fold)
plici- pleated, folded lengthwise, plicate
ploco- chapletted- (e.g. a whorl of follicles – *plococarpus*)
plococarpus with whorled fruits
plumarius -a -um, *plumatus -a -um* feathery, plumed, plumose
Plumbago Leaden, *plumbus* (Pliny's name refers to the flower colour)
plumbeus -a -um lead-coloured
Plumeria for Charles Plumier (1646–1704), French botanist
plumosus -a -um feathery, plumed
plur-, *pluri-* many-, several-
pluridens many-toothed
pluriflorus -a -um many-flowered
pluvialis -is -e, *pluviatilis -is -e* announcing rain, growing in rainy places
pneuma-, *pneumato-* air-, respiratory- (πνευμων)
pneumonanthe (us) lung-flower (the former use of marsh gentian, *Gentiana pneumonanthe* for respiratory disorders)

Poa Pasturage (the Greek name, ποα, for a fodder grass)
pocophorus -a -um fleece-bearing
poculiformis -is -e goblet-shaped (with upright limbs of the corolla)
pod-, podo-, -opus, -podius -a -um foot-, stalk, -foot (ποδος)
podagrarius -a -um, podagricus -a -um snare, of gout (used to treat gout)
Podalyra for Podalyrius, son of Aesculapius
podalyriaefolius -a -um with leaves resembling those of *Podalyra*
podeti- stalk-
-podioides -foot-like
-podius -a -um, podo-, -podus -a -um foot, stalk, ποδος
Podocarpus Foot-fruit (the characteristic shape of the fleshy fruit-stalks of some)
poecilo-, poikilo- variable-, variegated-, variously- (ποικιος)
poetarum, poeticus -a -um of poets (Greek gardens included games areas and theatres)
Poga from a vernacular name from Gabon, West Africa
pogon-, -pogon bearded-, -haired, -bearded, πωγων
poikilo- variable-, variegated-, spotted-
poissonii for M. Poisson (*c.* 1881), French botanist
polaris -is -e from the North polar region, of the North Pole
Polemonium for King Polemon of Pontus (the name used by Pliny)
poli-, polio- grey- (πολιος)
Polianthes Grey-flowered
polifolius -a -um grey-leaved, *Teucrium*-leaved
Poliothyrsis Greyish-panicle (the colour of the inflorescence)
politus -a -um elegant, polished
polius -a -um greyish-white (foliage)
pollacanthus -a -um flowering repeatedly
pollicaris -is -e as long as the end joint of the thumb (*pollex*), about one inch
pollinosus -a -um as though dusted with fine flour (pollen)
polonicus -a -um from Poland, Polish
poly- many- (πολυς)

polyacanthus -a -um many-spined
Polyalthia Many-healing (the supposed properties of the flowers)
polyanthemos, polyanthus -a -um many-flowered
Polycarpon Many-fruited (a name, πολυκαρπος, used by Hippocrates)
Polyceratocarpus Many-horned-fruits
polyedrus -a -um many-sided
Polygala Much-milk (Dioscorides' name, πολυγαλον, refers to the improved lactation in cattle fed on milkworts)
polygamus -a -um the flowers having various combinations of the reproductive structures
Polygonatum Many-knees (the structure of the rhizome)
Polygonum Many-joints (the swollen stem nodes)
polygyrus -a -um twining
polymorphus -a -um variable, of many forms
polypodioides resembling *Polypodium*
Polypodium Many-feet (πολυς), possibly from the rhizome growth pattern
Polypogon Many-bearded
Polyporus Many-pored
Polyscias Many-umbelled
Polystichum Many-rows (the arrangement of the sori on the fronds)
polystomus -a -um with many suckers or haustoria (στομα, mouth)
Polytrichum Many-hairs, the surface covering of the calyptra
pomaceus -a -um pome-bearing, apple-like (*pomum*, apple)
pomeridians, pomeridianus -a -um of the afternoon, p.m. (afternoon-flowering)
pomi- apple-like-
pomifer -era -erum apple-bearing, pome-bearing
pomponius -a -um of great splendour, pompous, having a top-knot or pompon
Poncirus from the French name, poncire, for a kind of *Citrus*
ponderosus -a -um heavy, large, ponderous

Pontederia for Guillo Pontedera (1688–1757), Professor of Botany at Padua
ponticus -a -um of the Black Sea's southern area, Pontica
pontophilus -a -um living in the deep sea
poophilus -a -um meadow-loving
populeus -a -um blackish-green (colour of leaves of *Populus nigra*)
populifolius -a -um poplar-leaved
populneus -a -um (populnaeus) poplar-like, related to Populus
Populus the ancient name for poplar, *arbor populi* 'tree of the people'
por- passage-, pore-
porcatus -a -um ridged
porcinus -a -um of pigs
porophilus -a -um loving soft stony ground
porophyllus -a -um having (or appearing to have) holes in the leaves
porosus -a -um with holes or pores
porphyreus -a -um, porphyrion warm-reddish-purple (πορφυρεος)
porra-, porri- leek-like-, leek-, *porrum*-like-
porrectus -a -um spreading, long, protracted
porrifolius -a -um leek-leaved
porrigens spreading (*porrigo* to spread)
porrigentiformis -is -e porrigens-like (the leaf-margin teeth point outwards and forwards)
porrum a Latin name used for various *Allium* species
porulus -a -um somewhat porous
poscharskyanus -a -um for Gustav Adolf Poscharsky, one time garden inspector in Laubegast, Dresden
postmeridianus -a -um of the afternoon
Portenschlagia, portenschlagianus -a -um for Franz Elder von Portenschlag-Ledermeyer (1772–1822), Austrian botanist
portensis -is -e from Oporto, Portugal
portlandicus -a -um from the Portland area
portoricensis -is -e from Puerto Rico, West Indies
portula abbreviated form of *Portulaca*
Portulaca Milk-carrier (a name in Pliny)

portulaceus -a -um   *Portulaca*-like
post-   behind-, after-, later-
posticus -a -um   turned outwards from the axis, extrorse
potam-, potamo-   watercourse-, of watercourses-, river- (ποταμος)
Potamogeton   Watercourse-neighbour, ποταμος–γειτων (the habitat)
potamophilus -a -um   river-loving
potaninii   for Grigori Nicholaevich Potanin (1835–1920), Russian explorer
potatorum   of drinkers (used for fermentation)
Potentilla   Quite-powerful (as a medicinal herb)
Poterium   Drinking-cup (Dioscorides' name, ποτιρριον, for another plant)
poukhanensis -is -e   from Pouk Han, Korea
-pous   -foot, -stalk, -stalked, πους, ποδος
prae-, pre-   before-, in front-
praealtus -a -um   very tall or high, outstanding
praecox   earlier than most of its genus, early developing, precocious
praegeri   for the wife of Robert Praeger (1865), Dublin librarian and writer on *Sedum* and *Sempervivum*, etc.
praegnans   full, swollen, pregnant (-looking)
praemorsus -a -um   as if nibbled at the tip
praepinguis -is -e   very rich
praerosus -a -um   appearing to have been gnawed off
praeruptorum   of rough places (living on screes)
praestans   pre-eminent, excelling
praeteritus -a -um   of the past
praetermissus -a -um   overlooked, omitted
praetextus -a -um   bordered
praeustus -a -um   appearing to have been scorched
praevernus -a -um   before spring, early, prevernal
prasinus -a -um, prasus -a -um   leek-green, leek-like (for various *Allium* species)
pratensis -is -e   of the meadows

pratericolus -a -um, praticolus -a -um of meadows, living in grassy places

pravissimus -a -um very crooked

precatorius -a -um relating to prayer, of petitions (*Abrus precatorius* rosary beads)

prehensilis -is -e grasping (flowers pollinated by insects that grasp the style or stamens)

prenans drooping

Prenanthes Drooping-flower (the nodding flowers)

preptus -a -um eminent

primitivus -a -um typical (in contrast to hybrids and varieties)

Primula Little-firstling (spring-flowering)

primulinus -a -um primrose-coloured, *Primula*-like

primuloides resembling *Primula*

princeps, principis -is -e most distinguished, first, princely

Prinsepia for James Prinsep (1778–1840), meteorologist of the Asiatic Society of Benghal

prio-, priono- serrated-, saw-toothed-, πριων

priochilus -a -um saw-lipped

Prionium Saw (πριονιον), the leaf-margins

prismati-, prismaticus -a -um prism-, prism-like-

pro- forwards-, for-, instead of-, before-

proboscidius -a -um trunk-like (the spadix of the mouse plant *Arisarum proboscidium*)

Proboscoidea Trunk-like (προβοσκις, for obtaining food)

proboscoides, proboscoideus -a -um snout-like, trunk-like

procerus -a -um very tall

procumbens lying flat on the ground, creeping forwards, procumbent

procurrens spreading below ground, running forwards

prodigiosus -a -um wonderful, marvellous, prodigious

productus -a -um stretched out, extended, produced

profusus -a -um very abundant, profuse

prolepticus -a -um developing early, precocious

prolifer -era -erum producing offsets or young plantlets or bunched growth, proliferous
prolificus -a -um very fruitful
prominens outstanding
pronatus -a -um, pronus -a -um lying flat, with a forward tilt
propaguliferus -a -um prolific, multiplying by vegetative propagules
propendens, propensus -a -um hanging down
propinquus -a -um closely allied, of near relationship, related
pros- near-, in addition-, also-
proso-, prostho- towards-, to the front-, before-
Prostanthera Appendaged-anther
prostratus -a -um lying flat but not rooting, prostrate
Protea for Proteus (the sea god's versatility in changing form)
proter-, protero-, proto- first-
Protomegabaria Former-*Megabarya* (relationship to the genus *Megabarya*)
protruberans bulging out
protrusus -a -um protruding
provincialis -is -e from Provence, France
pruhonicus -a -um from Pruhonice, former Czechoslovakia
pruinatus -a -um, pruinosus -a -um powdered, with a hoary bloom as though frosted over
Prunella (Brunella) from the German name, die Braune, for quinsy for which it was used as a cure
pruni- plum-like, plum-, *Prunus-*
Prunus the ancient Latin name for a plum tree
pruriens irritant, stinging, itch-causing (hairs on the fruits of *Mucuna pruriens*)
przewalskii for Nicholas Przewalski
psamma-, psammo- sand-
Psamma Strand-dweller (an old generic name for marram grass refers to its habitat)
pseud-, pseudo- sham-, false- (ψευδο-)

pseudacacia false *Acacia* (the similar appearance of *Robinia pseudacacia*)

Pseudagrostistachys False-grasslike-spike (refers to the short axillary racemes)

Pseuderanthemum False-*Eranthemum*

Pseudolarix False-*Larix*

Pseudopanax False-*Panax*

pseudosecalinus -a -um false *Bromus secalinus*

Pseudotsuga False-*Tsuga*

Psidium a Greek name, ψιδιον, formerly for the pomegranate (similarity of the fruits)

psilo- slender-, smooth-, bare- (ψιλος)

psilostemon with slender- or naked-stamens

Psilotum Hairless

psittacinus -a -um parrot-like (contrasted colouration)

psittacorum of parrots

Psoralea Manged (the dot-marked vegetative parts)

Psyche Love (one of the Dryad nymphs, ψυκηε)

psychodes, psycodes butterfly-like, Psyche was a Dryad nymph married to Cupid

Psylliostachys Bare-spike (ψιλος)

psyllium of fleas (from a Greek name, refers to the resemblance of the seed to a flea, ψυλλα)

ptarmicoides ptarmica-like, resembling *Achillea ptarmica*

ptarmica causing sneezes (an old, onomatopoeic, generic name πταρμικη)

Ptelea the ancient Greek name, πτελεα, for elm (transferred for the similarity of the fruit)

Pteleopsis *Ptelea*-like (resembling the hop-tree)

pteno- deciduous- (πτηνο-)

ptera-, ptero-, -pteris -is -e, ptery- with a wing- (πτερον), winged-

pteranthus -a -um with winged flowers

Pteridium Small-fern, πτεριδιον

Pteris Feathery (the Greek name, πτερις, for a fern)

Pterocarya Winged-nut (the winged fruits of most)
Pterocephalus Winged-head (appearance of the senescent flower-heads)
Pterostyrax Winged-*Styrax* (one species has winged fruits)
-pterus -a -um -winged
-pterygius -a -um -winged (πτερυγιον)
Pterygota Winged (the *Acer*-fruit-like seed)
ptilo- feathery- (πτιλον)
ptycho- folded- (πτυξ,)
Ptychopyxis Folded-capsule
puberulus -a -um downy
pubens full-grown, pubescent, juicy
pubescens softly hairy, covered with down, downy, pubescent
pubi-, pubigenus -a -um, pubigerus -a -um hairy
pubibundus -a -um with much downy hair
Pubilaria Hairy (the clothing of fibrous leaf remains on the rhizome)
Puccinellia for B. Puccinelli (1808–1850), Italian botanist of Lucca
puddum from a Hindi name for a cherry
puderosus -a -um very bashful
pudicus -a -um retiring, modest, bashful
Puelia, puellii for Timothee Puel (1812–1890), French botanist of Paris
Pueraria for M.N. Puerari (1765–1845), Swiss Professor of Botany at Copenhagen
pugioniformis -is -e (us) dagger-shaped
pulchellus -a -um beautiful, pretty
pulcher -ra -rum beautiful, handsome, fair
pulcherrimus -a -um most beautiful, most handsome
Pulegium Flea-dispeller (a Latin plant name)
Pulicaria Fleabane; Latin name for a plant which wards off fleas (*pulicis*)
pulicaris -is -e of fleas (e.g. the shape of the fruits)
pullatus -a -um clothed in black, sad-looking
pullus -a -um raven-black, almost dead-black

Pulmonaria Lung-wort (the signature of the spotted leaves as indicative of efficacy in the treatment of respiratory disorders)
pulposus -a -um fleshy, pulpy
Pulsatilla Quiverer (Brunfels' name for the movement of the flowers in the wind)
pulverulentus -a -um covered with powder, powdery, dusty (*pulvis* dust)
pulviger -era -erum dusted, powdered
pulvinatus -a -um cushion-like, cushion-shaped, with pulivini
pumilus -a -um low, small, dwarf
punctati-, *puncti-*, *punctatus -a -um* with a pock-marked surface, spotted, punctate
puncticulatus -a -um, *puncticulosus -a -um* minutely spotted
punctilobulus -a -um dotted-lobed
pungens ending in a sharp point, pricking
Punica Carthaginian (from a name, *malum punicum*, in Pliny)
puniceus -a -um crimson, carmine-red
punici- pomegranate-like, *Punica-*
purga purgative (the officinal root, Jalap, of *Ipomoea purga*)
purgans, *purgus -a -um* purgative
purpurascens, *purpurescens* becoming purple
purpuratus -a -um empurpled, purplish
purpureus -a -um reddish-purple
purpurinus -a -um somewhat purplish
purpusii for either of the brothers J.A. and C.A. Purpus of Darmstadt
Purshia for F.T. Pursch (1774–1820), author of *Flora Americae septentrionalis*
purus -a -um clear, spotless, pure
-pus -foot
Puschkinia for Count Graffen Apollos Apollosovitsch Mussin-Puschkin (d. 1805), Russian phytochemist and plant collector in the Caucasus
pusillus -a -um insignificant, minute, very small, slender, weak
pustulatus -a -um as though covered with blisters

pustulosus -a -um pustuled, pimpled
puteorum of the pits
putens foetid, stinking
pycn-, pycno- close-, densely-, compact-, dense- (πυκνος)
Pycnanthus, pycnanthus -a -um Densely-flowered
pycnostachyus -a -um close-spiked
pygmaeus -a -um, pygmeus -a -um dwarf
Pyracantha Fire-thorn
pyracanthus -a -um fire-thorned (persistent irritation caused by the thorns)
pyramidalis -is -e, pyramidatus -a -um conical, pyramidal
pyraster an old generic name, pear-flowered
pyren-, pyreno- kernel-, stone-
pyrenaeus -a -um, pyrenaicus -a -um from the Pyrenees Mountains
Pyrethrum Fire (medicinal use in treating fevers)
pyri- pear-
pyriformis -is -e pear-shaped
pyro-, pyrro, pyrrho- fire-, πυρ, πυρρος
Pyrola Pear-like (compares the leaves with those of *Pyrus*)
pyropaeus -a -um flame-coloured
pyrophilus -a -um growing on burnt earth
Pyrularia Little-pear (allusion to shape of the fruit)
-pyrum -wheat
Pyrus from the ancient Latin name, *pirus*, for a pear tree
pyxidatus -a -um small-box-like, *pyxis*, with a lid (e.g. some stamens)

quad-, quadri- four-
quadrangularis -is -e, quadrangulatus -a -um with four angles, quadrangular
quadratus -a -um four-sided, square-stemmed
quadriauritus -a -um four-lobed, four-eared
quadrifidus -a -um divided into four, cut into four
quadrijugatus -a -um with four pairs of leaflets
quadriquetrus -a -um square-sided, four-sided

quamash from the North American Indian vernacular name for *Camassia* bulbs used as food

Quamoclit from the Mexican vernacular name for *Ipomaea quamoclit*, Indian Pink

quaquaversus -a -um growing in all directions

quarter- having four-

Quassia for the Surinamese slave, Quassi, who discovered the medicinal properties of *Quassia amara*, in 1730

quassioides resembling *Quassia*

quaternarius -a -um, *quaternatus -a -um* structures arranged in fours

quaternellus -a -um with four divisions, four-partite

querci-, *quercinus -a -um* oak-, oak-like, resembling *Quercus*

Quercus the old Latin name for an oak

-quetrus -a -um -angled, -acutely-angled, sided-

quichiotis chimerical, quixotic

quin-, *quini-*, *quinque-* five-

quinatus -a -um five-partite, divided into five (lobes), in fives

quincuncialis -is -e arranged like the spots on the five-side of a dice (quincunx) or aestivated with two members internal, two members external and the fifth half external and half internal, in five ranks

quinquangularis -is -e five-cornered, five-angled

quinquelocularis -is -e five-celled, five-locular (the ovary)

quinquevulnerus -a -um with five marks (e.g. on the corolla), five-wounded

Quisqualis Who? What? (from a Malay name 'udani' which Rumphius transliterated as Dutch 'hoedanig' for How? What?

quitensis -is -e from Quito, Ecuador

quintupli- five-

quintuplex in multiples of five

rabdo- see *rhabdo-* (ραβδος)

racem-, *racemi-* raceme-

racemi-, *racemosus -a -um* with flowers arranged in a raceme (see Fig. 2(*b*))
racemosus -a -um having racemose inflorescences
rache-, *-rachi* rachis-, -rachis, ραχις, backbone
rachimorphus -a -um backbone-like, with a zig-zag central axis (as in *Rottboellia*)
raddeanus -a -um for Gustav Ferdinand Richard Radde
radens rasping, scraping (the rough surface)
radialis -is -e radial, actinomorphic
radians, *radiatus -a -um* radiating outwards
radiatiformis -is -e with the ligulate florets increasing in length towards the outside of the capitulum
radicalis -is -e arising from a root or a crown
radicans with rooting stems
radicatus -a -um, *radicosus -a -um* with large, conspicuous or numerous roots
radiiflorus -a -um with radiating flowers or perianth segments
radinus -a -um slender
Radiola Radiating (the branches)
radiosus -a -um having many rays
radulus -a -um rough, rasping, like a rasp
ragusinus -a -um from Dubrovnik (*Ragusa*), former Yugoslavia
raffia, *roffia* see *Raphia*
rafflesianus -a -um for Sir Thomas Stamford Raffles (1781–1826), diplomat, naturalist and a founder of London Zoo
rakaiensis -is -e from the Rakai Valley, Canterbury, New Zealand
ramentaceus -a -um covered with scales (ramenta)
-rameus -a -um -branched
rami- branches-, of branches-, branching-
ramiflorus -a -um with flowers on the branches
Ramischia for F.X. Ramisch (1798–1859), Bohemian botanist
Ramonda for Louis François Elisabeth Ramond de Carbonnières (d. 1827), French botanist

ramosissimus -a -um greatly branched

ramosus -a -um branched

ramulosus -a -um twiggy

ranunculoides *Ranunculus*-like

Ranunculus Little-frog (the amphibious habit of many)

Raoulia for Edouard F.A. Raoul (1815–1852), French surgeon and writer on New Zealand plants

rapa an old Latin name for a turnip

rapaceus -a -um of turnips, *Rapa*-like

raphani- radish-, radish-like-

Raphanus the Latin name for a radish

Raphia from the Malagasy name for the fibres from *Raphia pedunculata*, or Needle (the sharply pointed fruit)

Raphiolepis Needle-scale (the subulate bracts)

Rapistrum Rape-flower (implies inferiority of wild mustard)

rapum-genistae rape-of-broom (parasite of *Sarothamnus*)

rapunculoides resembling *Rapunculus*, rampion-like

Rapunculus Little-turnip (the swollen roots)

rari- thin-, scattered-, loose-

rariflorus -a -um having scattered flowers

rarus -a -um uncommon, scattered, distinguished

Ravenala from the Madagascan name for the travellers' tree

ravidus -a -um, *ravus -a -um* tawny-grey-coloured

re- back-, again-, against-, repeated-

recedens retiring, disappearing

reclinatus -a -um drooping to the ground, reflexed, bent back, reclined

reclusus -a -um see *inclusus*

recognitus -a -um authentic, the true one

reconditus -a -um hidden, not conspicuous, concealed

rect-, *recti-* straight-

rectangularis -is -e rectangular

rectinervis -is -e, *rectinervius -a -um* straight-veined

rectus -a -um straight, upright, erect

recurvatus -a -um, recurvi-, recurvus -a -um curved backwards, recurved
recutitus -a -um skinned, circumcised (the appearance caused by the reflexed ray florets of the flower head)
redactus -a -um reduced, rendered fruitless
redivivus -a -um coming back to life, renewed (perennial habit or reviving after drought)
reductus -a -um drawn back, reduced
reflexus -a -um bent back upon itself, reflexed
refractus -a -um abruptly bent backwards, broken(-looking)
regalis -is -e outstanding, kingly, royal, regal
regerminans regenerating
regerminatus -a -um re-sprouting
regina, reginae queen, of the queen
regis-jubae King-Juba, who was a king of Numidia
regius -a -um splendid, royal, kingly
regma- fracture-
regmacarpius -a -um with a schizocarp breaking into cocci
regularis -is -e uniform, actinomorphic
regulus goldcrest
rehderi either for Jacob Heinrich Rehder (1790–1852), of Moscow; or for Alfred Rehder (1863–1949), see below
Rehderodendron for Alfred Rehder of the Arnold Arboretum, author of the standard work, *Manual of Cultivated Trees and Shrubs*
religiosus -a -um sacred, venerated, of religious rites (the Buddha is reputed to have received enlightenment beneath the bo tree, *Ficus religiosa*)
remotus -a -um scattered (e.g. the flowers on the stalk)
renarius -a -um, reniformis -is -e kidney-shaped, reniform
repandens, repandus -a -um with a slightly wavy margin, repand
repens creeping (stoloniferous)
replicatus -a -um double-pleated, doubled down
reptans creeping
reptatrix creeping-rooted
resectus -a -um shredded, cut off

resedi- Reseda-

Reseda Soother (the name, *resedo*, in Pliny refers to its use in treating bruises)

resinifer -era -erum, *resinosus -a -um* producing resin, resinous

restibilis -is -e perennial

resupinatus -a -um inverted (e.g. those orchids with twisted ovaries), resupine

ret-, *reti-* net-

retatus -a -um netted, net-like

reticulatus -a -um netted, conspicuously net-veined, reticulate

retortus -a -um twisted back

retractus -a -um drawn backwards

retro- back-, behind-, backwards-

retroflexus -a -um, *retrofractus -a -um* turned backwards or downwards

retrorsus -a -um curved backwards and downwards

retusus -a -um blunt with a shallow notch at the tip (e.g. leaves, see Fig.7 (*f*)), retuse

reversus -a -um reversed

revolutus -a -um rolled back, rolled out and under (e.g. leaf margin), revolute

rex king

rhabdo- rod-like, rod-, ραβδος

Rhabdothamnus Rod-bush (much branched)

rhabdotus -a -um striped

rhache-, *-rhachis -is -e* rachis-, -rachis (ραχις, backbone)

rhaeticus -a -um from the Rhaetian Alps of the Swiss/Austrian border

rhaga-, *-rhagius -a -um* -torn, -rent (ραγας a fissure)

Rhamnella Little-*Rhamnus*

Rhamnus an ancient name, ραμνος, for various prickly shrubs

rhaphi-, *rhaphio-* needle-like-, needle- (ραφις)

Raphiolepis Needle-scaled

rhaponticus -a -um rhubarb from the Black Sea area (ρα of Dioscorides, with *pontus*) *rhapto-* stitched- (ραπτος)

Rhaptopetalum Seamed-petals (the valvate corolla)

Rheum from a Persian name, rha, for rhubarb

rhin-, rhino- nose- (ρις, ρινος)

Rhinanthus Nose-flower

rhipi- fan-shaped- (ριψ, ριπις)

Rhipsalis Wickerwork-like (the slender twining stems)

rhiz-, rhizo- root- (ριζα)

Rhizanthemum Root-flower (Malaysian parasitic plant)

rhizophyllus -a -um root-leaved (the leaves form roots)

Rhizophora Root-carrier (the long-arched prop-roots)

-rhizus -a -um -rooted, -root (ριζα)

rhod-, rhodo- rose-, rosy-, red- (ροδον)

Rhodanthemum Red-flower (*Chrysanthemum*)

rhodantherus -a -um with red stamens

rhodanthus -a -um rose-flowered

rhodensis -is -e, rhodius -a -um from the Aegean Island of Rhodes

Rhodochiton Red-cloak (the large calyx)

Rhododendron (um) Rose-tree (an ancient Greek name, ροδοδενδρον, used for *Nerium oleander*)

rhodopaeus -a -um, rhodopensis -is -e from Rhodope Mountains, Bulgaria

rhodophthalmus -a -um red-eyed

Rhodothamnus Rose-shrub (the flower colours)

Rhodotypos (-us) Rose-pattern (floral resemblance)

-rhoea -flowing (the sap or an exudate)

rhoeas the old generic name of the field poppy

Rhoeo Flowing (etymology uncertain but could refer to the mucilaginous sap)

Rhoicissus Pomegranate-ivy (ροια)

rhombi-, rhombicus -a -um, rhomboidalis -is -e, rhomboidosus -a -um diamond-shaped, rhombic

Rhombiphyllum Rhomboid-leaf (*R. rhomboideum* tells one little more!)

rhopalo- club-, cudgel- (ροπαλον)
rhumicus -a -um from the River Rhume area, W Germany
Rhus from an ancient Greek name for a sumach
rhynch-, rhyncho- beak- (ρυγχος)
Rhynchanthus Beak-flower (the protruding, keeled filament)
Rhynchelytrum (on) Beaked-sheath (the shape of the glumes)
Rhynchosia Beak (the shape of the keel petals)
Rhynchosinapis Beaked-*Sinapis*
Rhynchospora Beaked-seed
rhyti-, rhytido- wrinkled-, ρυτιδος
rhytidophyllus -a -um with wrinkled leaves
rhyzo-, -rhyzus -a -um root-, -rooted
Ribes from the Persian, ribas, for an acid-tasting *Rheum*
Ricinodendron *Ricinus*-like-tree (a similarity of the foliage)
Ricinus Tick (the appearance of the seeds)
rigens stiffening, rigid
rigensis -is -e from Riga (Latvia), on the Baltic
rigescens of a stiff texture
rigidus -a -um stiff, inflexible
rimosus -a -um with a cracked surface, furrowed
ringens with a two-lipped mouth, gaping
ringo from the Japanese vernacular name for *Malus ringo*
Rinorea from a Guyanese vernacular name
riparius -a -um of the banks of streams and rivers
ritro a southern European name for *Echinops ritro*
rivalis -is -e of brooksides and streamlets
Rivina, riviniana for August Quirinus Rivinus (1652–1722), former Professor of Botany at Leipzig
rivularis -is -e waterside, of the rivers
rivulosus -a -um with sinuate marking or grooves
robbiae for Mary Anne Robb (1829–1912), who reputedly smuggled *Euphorbia amygdaloides* ssp. *robbiae*, Mrs Robb's hat, from Turkey, in a hatbox
robertianus -a -um of Robert (which Robert is uncertain)

Robinia for Jean Robin (1550–1629) and Vesparian Robin (1579–1600), herbalists and gardeners to Henry VI of France
robur oak timber, strong, hard
robustus -a -um strong-growing, robust
Rodgersia for Rear Admiral John Rodgers (1812–1882), expedition commander of the US Navy
Roemeria for Johann Jacob Römer (1763–1819), Swiss botanist
romanus -a -um of Rome, Roman
Romneya for Rev. T. Romney Robinson (1792–1882), Irish astronomer
Romulea for Romulus, founder of Rome
roribaccus -a -um dewberry
roridus -a -um with apparently minute blisters all over the surface, bedewed
Rorippa from an old Saxon name
rorulentus -a -um dewy
Rosa the Latin name for various roses
rosaceus -a -um looking or coloured like a rose
rosae-, rosi-, roseus -a -um rose-like, rose-coloured
Roscoea for William Roscoe (1753–1831), founder of the Liverpool Botanic Garden
roseolus -a -um pink or pinkish
rosmarini- *Rosemarinus-*
Rosmarinus Sea-dew (an ancient Latin name)
rostellatus -a -um with a small beak, beaked
rostratus -a -um with a long straight hard point, beaked, rostrate
rostri-, rostris -is -e, rostrus -a -um nose-, beak-like
Rosularia Little-rose (the leaf rosettes)
rosularis -is -e, rosulatus -a -um with leaf rosettes
rotatus -a -um flat and circular, wheel-shaped
rotundi-, rotundus -a -um rounded in outline or at the apex, spherical
Roystonia for General Roy Stone (1836–1905), American soldier
-rrhizus -a -um -rooted
rubellinus -a -um, rubellus -a -um reddish

rubens blushed with red, ruddy
ruber, rubra, rubrum, rubri-, rubro- red
rubescens, rubidus -a -um turning red, reddening
Rubia a name in Pliny for madder
rubicundus -a -um ruddy, reddened
rubiginosus -a -um, rubrus -a -um rusty-red
Rubus the ancient Latin name for brambles
Rudbeckia for Linnaeus' mentor Olof O. Rudbeck (1660–1740) and his son J.O. Rudbeck (1711–1790)
rudentus -a -um cabled, rope-like
ruderalis -is -e of waste places, of rubbish tips
rudis -is -e untilled, rough, wild
rudiusculus -a -um wildish
rufescens, rufidus -a -um becoming reddish, turning red
rufi- red-, reddish-
rufinus -a -um red
rufus -a -um, -rufus rusty (-haired), pale- or reddish-brown, red (in general)
rugosus -a -um wrinkled, rugose (e.g. leaf surfaces)
rugulosus -a -um somewhat wrinkled, with small wrinkles
Rumex a name in Pliny for sorrel
rumici- dock-like-
ruminatus -a -um thoroughly mingled, as if chewed
runcinatus -a -um with sharp retrorse teeth (leaf margins), saw-toothed with the fine tips pointing to the base, runcinate
rupester -tris -tre, rupicola of rock, of rocky places
rupi-, rupri- of rocks-, of rocky places-
rupifragus -a -um growing in rock crevices, rock-cracking
Ruppia (Ruppa) for H.B. Ruppius (1688–1719), German botanist
rupti- interrupted-, broken-
ruralis -is -e of country places, rural
rurivagus -a -um of country roads, country wandering
Ruschia for E. Rusch, South African farmer
rusci- box holly-like, butcher's broom-like, resembling *Ruscus*

Ruscus an old name for a prickly plant

russatus -a -um reddened, russet

Russelia for Dr Alexander Russel, author of *Natural History of Aleppo* (1775)

russotinctus -a -um red-tinged

rusticanus -a -um, *rusticus -a -um* of wild places, of the countryside, rustic

Ruta the ancient Latin name for rue

ruta-baga from a Swedish name

ruta-muraria rue-of-the-wall, a name used in Brunfels

ruthenicus -a -um from Ruthenia, Carpathian Russia

rutilans, *rutilus -a -um* deep bright glowing red, orange, or yellow

rytidi-, *rytido-* wrinkled- (ρυτις, ρυτιδος)

rytidocarpus -a -um wrinkled-fruit

sabatius -a -um from Capo di Noli, Riviera di Ponente, Italy

sabaudus -a -um from Savoy (*Sabaudia*), SE France

Sabbatia for L. Sabbati (*c.*1745), Italian botanist

sabbatius -a -um from Savona, NW Italy

sabdariffa from a West Indian name

Sabia from its Benghali name, sabja-lat

sabina from the Latin name, *herba sabina*, for Savin, *Juniperus sabina*

sabinianus -a -um for Joseph Sabine (1770–1837), founder of the Horticultural Society of London

sabrinae from the River Severn (*Sabrina*)

sabulicolus -a -um, *sabulus -a -um* living in sandy places, sand-dweller

sabulosus -a -um full of sand, of sandy ground

sacc- sac-, pouch- (σακκος)

saccatus -a -um bag-shaped, pouched, saccate

saccharatus -a -um with a scattered white coating, sugared, sweet-tasting

sacchariferus -a -um sugar-producing, bearing sugar (σακχαρον)

saccharinus -a -um, *saccharus -a -um* sweet, sugary

Saccharum Sugar (from the Latin name)

saccifer -era -erum having a hollowed part, pouch-bearing, bag-bearing
sachalinensis -is -e from Sakhalin Island, E Siberia
Sacoglottis Pouch-tongue (the anthers dehisce through basal pouch-like extensions)
sacrorum of sacred places, of temples, sacred (former ritual use)
saepium of hedges
Sagina Fodder (the virtue of a former included species, spurrey)
sagittalis -is -e, sagittatus -a -um, sagitti- (saggitatus) arrow-shaped, sagittate (see Fig. 6(*c*))
Sagittaria Arrowhead, *sagitta* (the shape of the leaf-blades re-emphasized in *Sagittaria sagittifolia*)
sago yielding the large starch grains, from Malay, sagu
Saintpaulia for Baron Walter von Saint Paul-Illaire (1860–1910), who discovered *S. ionantha*
Salaxis an unexplained name by Salisbury
salebrosus -a -um rough
salicarius -a -um, salicinus -a -um willow-like, resembling *Salix*
salice-, salici- willow-like, willow-
salicetorum of willow thickets
Salicornia Salt-horn (refers to the habitat and the form of the shoots)
saliens projecting forward
salignus -a -um of willow-like appearance, willowy, resembling *Salix*
salinus -a -um of saline habitats, halophytic
salisburgensis -is -e from Salzburg, Austria
Salix the Latin name for willows
salmoneus -a -um salmon-coloured, pink with a touch of yellow (in mythology, the son of Aeolus, punished for imitating lightning)
salpi- trumpet-, σαλπιγξ
Salpichroa Tube-of-skin (the flower), χρως, skin
Salpiglossis Trumpet-tongue (the shape of the style)
salsuginosus -a -um of habitats inundated by salt-water, of salt-marshes
Salsola Salt (the taste and the habitat)

salsus -a -um living in saline habitats
saltatorius -a -um dancing
saltitans jumping (heat-sensitive larvae of *Cydia saltitans* in seed of the Mexican jumping bean *Sebastiana* cause it to jump)
saltuum of glades, woodlands or ravines
salutaris -is -e healing, beneficial
salvi-, salvii- sage-like-, resembling *Salvia*
Salvia Healer (the old Latin name for sage with medicinal properties)
Salvinia for Professor Antonio Maria Salvini (1633–1722), Italian botanist and Greek scholar
salviodorus -a -um sage-scented
saman, Samanea from a South American name for *Pithecolobium saman*
samaroideus -a -um with samara-like fruits
sambac from the Arabic name for *Jasminum sambac*
sambuci-, sambucinus -a -um elder-like, resembling *Sambucus*
Sambucus from the Latin name for the elder tree
samius -a -um from the Isle of Samos, Greece
Samolus from a Celtic Druidic name
sanctus -a -um holy, sacred, chaste
sanderianus -a -um from the Sander Nursery
Sandersonia for John Sanderson (d. 1881), secretary of the Horticultural Society of Natal
sanguinalis -is -e, sanguineus -a -um, sanguineolentus -a -um blood-red, bloody
Sanguisorba Blood-stauncher (has styptic property)
Sanicula Little-healer (its medicinal property)
Sanseveria for Prince Raimond de Sansgrio of Sanseviero (1710–1771)
santalinus -a -um sandal-wood
Santalum from the Persian name, shandal, for sandalwood
Santolina Holy-flax
Sanvitalia for the San Vitali family of Parma

sap-, *sapon-* sap-, sweet-tasting-, soapy- (*sapa*, plant-juice; *sapo*, soap)
sapidus -a -um pleasant-tasted, flavoursome, savoury
sapientium of the wise, of man (implies superiority compared with *troglodytarum*)
Sapindus Indian-soap (contraction of *sapo indicus*, from its use)
Sapium Soapy (refers to the sticky sap)
saponaceus -a -um, *saponarius -a -um* lather-forming, soapy
Saponaria Soap-like (lather-forming)
sapota from the Mexican name, cochil-zapotl, for chicle-tree; see also *zapota*
sappan from a Malayan vernacular name
sapphirinus -a -um sapphire-blue
saprio-, *sapro-* rotten-
Saraca from an Indian vernacular name
saracenicus -a -um, *sarracenicus -a -um* of the Saracens
Saracha for Isidore Saracha (1733–1803), a Benedictine monk who sent plants to the Madrid Royal Gardens
sarachoides resembling *Saracha*
sarc-, *sarco-* fleshy- (σαρξ, σαρκος)
Sarcobatus Fleshy-spiny-shrub
Sarcocephalus Fleshy-head (the head of fruits)
Sarcococca Fleshy-berry
sarcodes flesh-like (σαρκωδης)
sardensis -is -e from Sart (*Sardis*), Smyrna, Asia Minor
sardosus -a -um, *sardous -a -um* from Sardinia, Sardinian
Sargentodoxa Sargent's-glory, for Charles Sprague Sargent (1841–1927), Founder and Director of Arnold Arboretum, USA
sarmaticus -a -um from Sarmatia on the Russo-Polish border
sarmentaceus -a -um, *sarmentosus -a -um* with long slender stolons or runners
sarmentus -a -um twiggy
sarniensis -is -e from Guernsey (*Sarnia*), Channel Islands
saro- broom-like-

Sarothamnus Broom-shrub
Sarracenia for Michel Sarrazan (d. 1734), who introduced *S. purpurea* from Quebec
sarrachoides from a Brazilian name for *Solanum sarrachoides*
sartorii for Andria del Sarto (1486–1531), of tailors
Sasa the Japanese name for certain small bamboos
sasanqua from the Japanese name for the tea-oil producing *Camellia*
Sassafras from the Spanish name, salsafras, for a saxifrage with medicinal properties
sathro- humus-, decayed- (σαθρο)
sativus -a -um planted, cultivated, sown, not wild
saturativirens full-deep-green
Satureia, Satureja the Latin name in Pliny for a culinary herb, savory
saur-, sauro- lizard-like-, lizard- (σαυρα)
Sauromatum Lizard (σαυρος), the figuring on the inner surface of the spathe
Saussurea for the Swiss philosopher Horace Benedict de Saussure (1740–1799)
savannarum of savannas
saxatilis -is -e living in rocky places, of the rocks
Saxegothaea for Prince Albert of Saxe-Coburg-Gotha, consort of Queen Victoria
saxicolus -a -um rock-dwelling
Saxifraga Stone-breaker (lives in rock cracks and had a medicinal use for gallstones)
saxosus -a -um of rocky or stony places
scaber -ra -rum coarse, rough, scabrid (like sandpaper)
scaberulus -a -um roughish, somewhat rough
Scabiosa Scabies (former medicinal use as a treatment for the disease)
scabri- rough-, scabrid-
scabriusculus -a -um somewhat scabrid
scabrosus -a -um rather rough
scalariformis -is -e with ladder-like markings, ladder-like

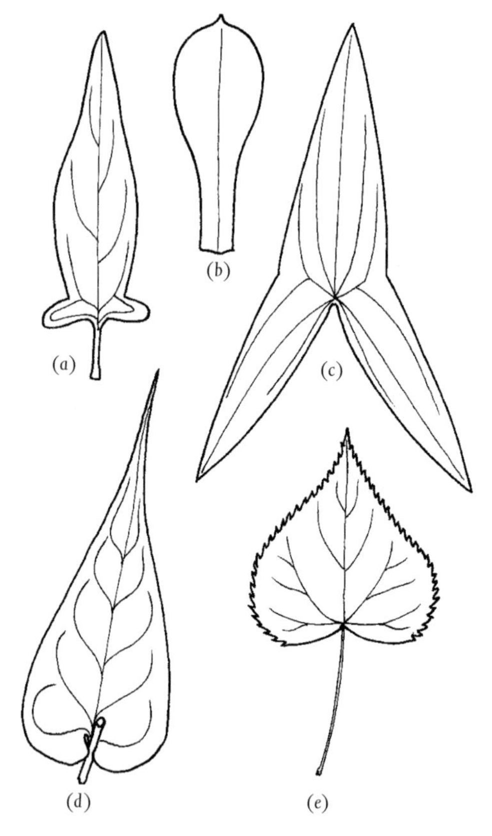

Fig. 6. More leaf shapes which provide specific epithets.
(*a*) Hastate (e.g. *Scutellaria hastifolia* L.) with auricled leaf-base;
(*b*) spathulate (e.g. *Sedum spathulifolium* Hook.);
(*c*) sagittate (e.g. *Sagittaria sagittifolia* L.) with pointed and divergent auricles;
(*d*) amplexicaul (e.g. *Polygonum amplexicaule* D.Don) with the basal lobes of the leaf clasping the stem;
(*e*) cordate (e.g. *Tilia cordata* Mill.) heart-shaped.

scalaris -is -e ladder-like
scalpellatus -a -um knife-like, cutting
scalpturatus -a -um engraved
scandens climbing
scandicus -a -um from Scandia, Scandinavian
Scandix ancient name, σκανδιξ, for shepherd's needle
scaphi-, *scapho-*, *scaphy-* boat-shaped-, bowl-shaped- (σκαφη)
Scaphopetalum Boat-shaped-petal
scapi-, *scapio-*, *-scapus -a -um* clear-stemmed-, scapose-, *scapus*
scapiger -era -erum scape-bearing
scaposus -a -um with scapes or leafless flowering stems
-scapus -a -um -peduncled, -stalked, -scaped, -scapose
scardicus -a -um from Sar Planina (*Scardia*), former Yugoslavia
scariola (serriola) endive-like, of salads
scariosus -a -um shrivelled, thin, not green, membranous, scarious
scarlatinus -a -um bright-red
sceleratus -a -um of vile places, vicious, wicked (causes ulceration)
sceptrus -a -um of a sceptre
schafta a Caspian area vernacular name for *Silene schafta*
Schefflera for J.C. Scheffler of Danzig
Scheuchzeria for the brothers Jakob J. (1672–1733) and J. Scheuchzer (1684–1738), Professors of Botany at Zurich
schinseng from the Chinese name
Schinus from the Greek name for another mastic-producing plant (*Pistacia*)
Schisandra Divided-man (the cleft anthers of the type species)
schist-, *schismo-* divided-, cut-, cleft- (σχιστος)
schist-, *schisto-* stone-, *schistos*
schistaceus -a -um slate-coloured
schistosus -a -um slaty-coloured
schiz-, *schizo-* cut-, divided-, split- (σχιζω)
Schizaea Cut (the incised fan-shaped fronds)
Schizanthus Divided-flower (the lobes of the corolla in the poor man's orchid)

schizomerus -a -um splitting into parts
Schizophragma Cleft-wall (the opening of the capsule)
-schizus -a -um -cut, -divided
Schkuhria for Christian Schkuhr (1741–1811), German botanist
Schlumbergera for Federick Schlumberg, a field botanist
schoen-, schoeno- rush-like, resembling *Schoenus*
Schoenoplectus Rush-plait (σχοινος–πλεκτος)
schoenoprasus -a -um rush-like leek (the leaves)
Schoenus the old name for rush-like plants
scholaris -is -e of the school, restful
Schwenkia for J.T. Schwenk (1619–1671), Professor of Medicine at Jena
scia-, scio- shaded-, shade- (σκια)
sciadi-, sciado- canopy-, umbelled- (σκιαδος)
sciaphilus -a -um shade-loving
Sciadopitys Umbrella-pine (the leaves are crowded at the branch ends)
Scilla the ancient Greek name, σκιλλα, for the squill *Urginea maritima*
scilloides squill-like, resembling *Scilla*
Scindapsus an ancient Greek name for an ivy-like plant
scintillans sparkling, gleaming
sciophilus -a -um shade-loving
scipionum wand-like
scirpoideus -a -um rush-like, *Scirpus*-like
Scirpus the old name for a rush-like plant
scissilis -is -e splitting easily
scitulus -a -um neat, pretty
scitus -a -um fine
sciuroides curved and bushy, squirrel-tail-like (σκιουρος a squirrel)
sclarea clear (an old generic name for a *Salvia*, clary, used for eye lotions)
Scleranthus Hard-flower (texture of the perianth)
sclero- hard- (σκληρος)

Scleranthus Hard-flower (the calyx)
Scleropoa Hard-pasturage
scobi-, *scobiformis -is -e* resembling sawdust or shavings (*scobis* sawdust)
scobinatus -a -um rough as though rasped (*scobina* a rasp)
scole-, *scolo-* vermiform-, worm- (σκωληξ)
scolio- curved-, bent- (σκολιος)
scolopax of the woodcock (*Scolopax rusticola*)
Scolopendrium Dioscorides' name for the hart's tongue fern compares the numerous sori to the legs of a millipede (σχολοπενδρα)
Scolymus the ancient Greek name, σκολυμος, for the artichoke, *Scolymus hispanicus*, and its edible root
scolytus elm-bark beetle
scopa- twining-, twigged
scoparius -a -um, *scopellatus -a -um* broom-like (use for making besoms)
scopulinus -a -um twiggy
scopulorum of cliffs and rock faces
scorbiculatus -a -um with a scurfy texture (*scorbutus* scurvy)
Scordium Dioscorides' name, σκορδιον, for a plant with the smell of garlic
scorodonia an old generic name, σχοροδον, for garlic
scorodoprasum (scordoprasum) a name, σκορδοπρασον, used by Dioscorides for a garlic-like leek (has intermediate features)
scorpioidalis -is -e, *scorpioideus -a -um* coiled like the tail of a scorpion (e.g. the axis of an inflorescence)
scorpioides (scorpoides) curved like a scorpion's tail (see Fig. 3(*c*)), scorpion-like
Scorpiurus Dioscorides' name for the coiled fruit of *Scorpiurus sulcata*
scorteus -a -um leathery
Scorzonera derivation uncertain but generally thought to refer to use as an antifebrile in snakebite (Italian, scorzone)
scot-, *scoto-* of the dark-, darkness- (σκοτος)

scoticus -a -um from Scotland, Scottish

scotinus -a -um dusky, dark

scotophilus -a -um dark-loving (e.g. subterraneous chemotrophic organisms)

Scottellia for G.F. Scott-Elliot, boundary commissioner and plant collector in Sierra Leone

scottianus -a -um for Munro B. Scott or Robert Scott (1757–1808), of Dublin

scrinaceus -a -um with lidded cup-like fruits (as in *Lecythis*)

scriptus -a -um marked with lines which suggest writing

scrobiculatus -a -um with small depressions or grooves, pitted (*scrobis* a ditch)

Scrophularia Scrophula (signature of the glands on the corolla), many plants were used to treat the 'King's disease'

scrotiformis -is -e shaped like a small double bag, pouch-shaped

sculptus -a -um carved

scutatus -a -um like a small round shield or buckler

Scutellaria Dish (the depression of the fruiting calyx)

scutellaris -is -e, scutellatus -a -um shield-shaped, platter-like

scutiformis -is -e buckler-shaped

scypho-, -scyphus -a -um cup-, beaker-, goblet-

Scyphocephalium Goblet-headed (the inflorescences contain up to three heads each of numerous flowers)

scyt-, scyto- leathery-

Scytanthus Leathery-flowered (part of the adaptation to attract coprozoic pollinators)

Scytonema Thong-like (leathery filaments)

scytophyllus -a -um leathery-leaved

se- apart-, without-, out-

sebaceus -a -um, sebifer -era -erum tallow-bearing, producing wax

sebosus -a -um full of wax

Secale the Latin name for a grain like rye

secalinus -a -um rye-like, resembling *Secale*

sechellarus -a -um from the Seychelles, Indian Ocean

seclusus -a -um hidden, secluded
sectilis -is -e as though cut into portions
sectus -a -um, -sect, -sectus -a -um cut to the base, -divided, -partite
secundi-, secundus -a -um turned-, one-sided (as when flowers are all to one side of an inflorescence), secund
secundiflorus -a -um with the flowers all facing one direction, secund-flowered
Securidaca Axe (from the shape of the fruits)
securiger -era -erum axe-bearing (the shape of some organ)
Securinega Axe-refuser (the hardness of the timber of some species)
sedi-, sedoides (sedioides) stonecrop-like, resembling *Sedum*
Sedum a name, *sedo*, in Pliny (refers to the plant's 'sitting' on rocks, etc. in the case of cushion species)
segetalis -is -e, segetus -a -um of the cornfields, growing amongst crops
segregatus -a -um a component separated from a superspecies
seiro- rope-like, rope, σειρα
sejugus -a -um with six leaflets
Selaginella a diminutive of *selago* (see below)
selaginoides clubmoss-like, resembling *Selaginella*
selago the name in Pliny for *Lycopodium*, from the Celtic name for a Druidic plant, *Juniperus sabina*
seleni-, seleno- moon-
Selinum the name in Homer for a celery-like plant with lustrous petals (relates etymologically with *Silaum* and *Silaus*)
sellaeformis -is -e, selliformis -is -e saddle-shaped, with both sides hanging down (e.g. of leaves)
selligerus -a -um saddled, saddle-bearing
selloi, sellovianus -a -um, sellowii, seloanus -a -um for Friedrich Sellow (Sello) (1789–1831), German botanist
semestris -is -e half-yearly, of a half year
semi- half-
semidecandrus -a -um with (about) five stamens (*Tibouchina semidecandra* has ten stamens but five have yellow anthers and the other five form a self-coloured platform for visiting pollinators)

semilunatus -a -um half-moon-shaped
semipersistens half-persistent
semiteres half-cylindrical
semper- always-, ever-
semperflorens ever-flowering, with a long flowering season
sempervirens always green
sempervivoides, sempervivus -a -um houseleek-like, resembling *Sempervivum*
Sempervivum Always-alive, never-die
senanensis -is -e from Senan, China
senarius -a -um six-partite
Senecio Old-man (the name in Pliny refers to the grey hairiness as soon as fruiting commences)
senecioides (senecoides) groundsel-like, *Senecio-*
senescens ageing, turning hoary with whitish hairs
seni- six-, six-each-
senifolius -a -um six-leafleted
senilis -is -e aged, grey-haired
Senna from the Arabic name
sensibilis -is -e, sensitivus -a -um sensitive to a stimulus, irritable
senticosus -a -um thorny, full of thorns
sepal-, -sepalus -a -um sepal-, -sepalled (σκεπη)
sepiaceus -a -um dark-clear-brown, sepia-coloured (*sepia*, cuttlefish)
sepiarius -a -um, sepius -a -um growing in hedges, of hedges
sepincola hedge-dweller, inhabitant of hedges (*sepes*, a hedge)
sept-, septem- seven-
septalis -is -e of September (flowering or fruiting)
septi-, septatus -a -um having partitions, septate
septemfidus -a -um with seven divisions, seven-cut
septentrionalis -is -e of the north, northern
septifragus -a -um having a capsule whose valves break away from the partitions
septupli- seven-
sepulchralis -is -e of tombs, of graveyards

sepultus -a -um buried

Sequoia for the North American Indian, Sequoiah (1770–1843), who invented the Cherokee alphabet

Sequoiadendron *Sequoia*-tree (resemblance in size)

serapias an ancient name, σεραπιας, for an orchid

seri-, serici-, sericans, sericeus -a -um silky-hairy (sometimes implying Chinese)

serialis -is -e, seriatus -a -um with transverse or longitudinal rows

sericatus -a -um silken, σηρικος

sericifer -era -erum, sericofer -era -erum silk-bearing

sericus -a -um from China (*Seres*)

-seris -potherb (σερις, σεριδος)

serissimus -a -um very late

serotinus -a -um of late season, autumnal (flowering or fruiting)

serpens, serpentarius -a -um, serpentinus -a -um creeping, serpentine

serpentini of (growing on) serpentine rocks

serpyllum from an ancient name, ερπυλλος, for thyme

serpyllifolius -a -um thyme-leaved

serra-, -serras saw-, saw-like-, serrate-

Serrafalcus for the Duke of Serrafalco, archaeologist

serratifolius -a -um with markedly serrate leaves

Serratula Saw-tooth (the name in Pliny for betony)

serratus -a -um edged with forward-pointing teeth, serrate (see Fig. 4(*c*))

serriolus -a -um in ranks, of salad (from an old name for chicory)

serrulatus -a -um edged with small teeth, finely serrate, serrulate

Sesamum Hippocrates' name from the Semitic name

Sesbania from the Arabic name for *Sesbania sesban*

Seseli the ancient Greek name, σεσελι, σεσελις

Sesleria for Leonardo Sesler of Venice (d. 1785)

sesqui- one-and-one-half-

sesquipedalis -is -e about 18 inches long, the length of a foot and a half

sesseli-, sessilis -is -e attached without a distinct stalk, sessile *sessilifolius -a -um* leaves without petioles, sessile-leaved

seta-, setaceus -a -um, seti- with bristles or stiff hairs, bristly

Setaria Bristly (the hairs subtending the spikelets)

Setcreasea derivation obscure

seti-, setifer -era -erum, setiger -era -erum bearing bristles, bristly, *saeta-*

setispinus -a -um bristle-spined

setosus -a -um covered with bristles or stiff hairs

setuliformis -is -e thread-like

setulosus -a -um with fine bristles

-setus -a -um -bristled

sex- six-

sexangularis -is -e, sexangulus -a -um six-angled (stems)

sextupli- six-fold-

shallon from a Chinook North American Indian name

Shepherdia for John Shepherd (1764–1836), curator of Liverpool University Botanic Garden

Sherardia for William (1659–1728) or James Sherard

Shortia for Dr Charles W. Short (1794–1863), botanist of Kentucky, USA

siameus -a -um from Thailand (Siam)

Sibbaldia for Prof. Robert Sibbald (1643–1720) of Edinburgh

sibiraea, sibiricus -a -um from Siberia, Siberian

Sibthorpia for John Sibthorp (1758–1796), English botanist, and his son Humphrey

siccus -a -um dry

siculi- dagger-shaped-

siculus -a -um from Sicily, Sicilian

Sida from a Greek name for a water-lily

Sidalcea Like-*Sida*-and-*Alcea*

sidereus -a -um iron-hard

Sideritis the Greek name for plants used on wounds caused by iron weapons

Sideroxylon Iron-wood (the hard timber of the miraculous berry)

sieboldiana, sieboldii for Philipp Franz von Siebold (1796–1866), German physician and plant collector in Japan

Sieglingia for Prof. Siegling of Erfurt

Sigesbeckia (Siegesbeckia, Sigesbekia) for John George Sigesbeck (1686–1755), physician and botanist of Leipzig, director of the St Petersburg Botanic Garden

sigillatus -a -um with the surface marked with seal-like impressions

sigma-, sigmato- S-shaped (σιγμα, σιγματος)

sigmoideus -a -um S-shaped (σιγμα)

signatus -a -um well-marked, designated, signed

sikkimensis -is -e from Sikkim, Himalayas

silaifolia with narrow leaves as in pepper saxifrage, *Silaum silaus*

Silaum meaning uncertain, from Sila forest area of S Italy? (see *Selinum*)

silaus an old generic name in Pliny used for pepper saxifrage

Silene Theophrastus' name for *Viscaria*, another catchfly; Silenos was Bacchus' companion

sileni- *Silene*-like

siliceus -a -um growing on sand

silicicolus -a -um growing on siliceous soils

siliculosus -a -um having broad pods or capsules from which the two valves fall and leave a false membrane (*replum*) with the seeds (*silicula*)

siliquastrum (siliquastris) from the old Latin name for a pod-bearing tree, cylindric-podded

siliquosus -a -um having elongate pods or capsules as the last (*siliqua*)

-siliquus -a -um, siliqui- -podded

sillamontanus -a -um from Cerro de La Silla, South America

silvaticus -a -um, silvester -tris -tre of woodlands, wild, of the woods

Silybum Dioscorides' name, σιλυβον, for a thistle-like plant

Simarouba (Simaruba) from the Carib (South America) name for bitter damson

simensis -is -e   from Arabia, *Simenia*, Middle Eastern
Simethis   after the Oreadean nymph Simaethis
simia   of the ape or monkey (flower-shape or implying inferiority)
simili-, *similis -is -e*   resembling other species, like, the same, similar
similiflorus -a -um   having the flowers all alike (e.g. in an umbel)
simplex   undivided, entire, single
simplice-, *simplici-*   undivided, simple
simulans, *simulatus -a -um*   resembling, imitating, similar
Sinapis   the old name, σιναπι, used by Theophrastus for mustard
sinensis -is -e (chinensis -is -e)   from China, Chinese
singularis -is -e   unusual, singular
sinicus -a -um, *sino-*   of China, Chinese
sinistrorsus -a -um   turned to the left, twining clockwise upwards (as seen from outside)
Sinningia   for Wilhelm Sinning (1794–1874), head gardener at Bonn University
sino-   Chinese-
Sinofranchetia   for Adrien Franchet (1834–1900), French botanist who described many Chinese plants
Sinomenium   Chinese-moonseed (the curved stone of the fruit)
Sinowilsonia   for E.H. Wilson (1876–1930), introducer of Chinese plants
sinuatus -a -um, *sinuosus -a -um*, *sinuus -a -um*   waved, with a wavy margin (see Fig. 4(*e*)), sinuate
siphiliticus -a -um   see *syphiliticus*
sipho-, *-siphon*   tubular-, -pipe, -tube (σιφον)
sipyleus   flour or meal
sisalanus -a -um   from Sisal, Yucatan, Mexico
sisara   Dioscorides' name for a plant with an edible root
Sison   a name, σισον, used by Dioscorides
Sisymbrium   ancient Greek name, σισυμβριον, for various plants
Sisyrinchium   Theophrastus' name, σισυριγχιον, for an iris
sitchensis -is -e   from Sitka Island, Alaska
Sium   an old Greek name, σιον, for water plants

skapho- see *scapho-*

Skimmia from a Japanese name, Miyami shikimi

skio- see *scia-*, *scio-* (σκια)

skolio- see *scolio-* (σκολιος)

skoto- see *scoto-* (σκοτος)

smaragdinus of emerald, emerald-green (σμαραγδος)

Smeathmannia for Henry Smeathmann, who collected plants in Sierra Leone in 1771–1772

smilaci- *Smilax-*

Smilacina diminutive of *Smilax*

Smilax from an ancient Greek name

Smithia for James Edward Smith (1759–1828), writer on the Greek flora and founder of the Linnean Society

Smyrnium Myrrh-like (the fragrance), σμυρνα myrrh

soboliferus -a -um bearing soboles, producing vigorous shoots from the stem at ground level

socialis -is -e growing in colonies, in pure stands, dominant

soda alkaline (the calcined ash of *Salsola*)

sodomeus -a -um from the Dead Sea area (Sodom)

sol-, *solis -is -e* sun-, of the sun

solan-, *solani-* potato-, *Solanum*-like-

Solanum Comforter (an ancient Latin name in Pliny)

solaris -is -e of the sun, of sunny habitats

Soldanella Coin-shaped (the leaves) from the Italin, soldo, for a small coin

soleae- sandal-, *solea*

Soleirolia for Lieut. Henri Augustine Soleirol (b. 1792), collector of Corsican plants

solen-, *soleno-* box-, tube-, σωλην

Solenostemon Tube-stamens (their united filaments)

-solens -tubed

Solidago Uniter (used as a healing medicine)

solidifolius -a -um entire-leaved

solidus -a -um a coin, complete, entire, solid, dense, not hollow *solitarius -a -um* the only species (of a monotypic genus), with individuals growing in extreme isolation

Sollya for Richard Horsman Solly (1778–1858), plant anatomist

solstitialis -is -e of mid-summer (flowering-time)

soma-, -somus -a -um- -bodied (σωμα)

somnians asleep, sleeping

somnifer -era -erum sleep-inducing, sleep-bearing

sonchi- Sonchus-like-

Sonchus the Greek name, σογχος, for a thistle

sonorus -a -um loud, resonant

sophia wisdom, σοφια, (the use of flixweed in healing)

Sophora from an Arabic name for a pea-flowered tree

soporificus -a -um sleep-bringing, soporific

Sorbaria Mountain-ash-like (from the form of the leaves)

Sorbus the ancient Latin name for the service tree

sordidus -a -um neglected, dirty-looking

sorediiferus -a -um bearing soredia (σωρος); on lichens

soriferus -a -um bearing sori (σωρος); on ferns

Sorghum from the Italian name

sororius -a -um very closely related, sisterly

soulangiana, soulangii for Etienne Soulange-Bodin (1774–1846), French horticulturalist

spadiceus -a -um chestnut-brown, date-coloured, having a spadix (σπαδιξ)

spananthus -a -um having few flowers, sparsely-flowered (σπανιος)

Sparaxis Tear (σπαρασσο), the torn bracts

Sparganium Dioscorides' name, σπαργανιον, for bur-reed

Sparrmannia for Dr Andreas Sparrmann (1748–1820), of Cook's second voyage

sparsi-, sparsus -a -um scattered

Spartina old name for various plants used to make ropes

Spartium Binding or Broom (former uses for binding and sweeping)

spath-, spathi-, spatho-, spathus -a -um spatulate-, spathe- (as in arums)

spathaceus -a -um with a spathe-like structure (bracts or calyx)
Spathiphyllum Leafy-spathe
Spathodea Spathe-like (the calyx)
spathulatus -a -um, spathuli- shaped like a spoon (see Fig. 6(*b*))
spatiosus -a -um spacious, wide, large
spatulae-, -spatulatus spoon-, -spatulate
speciosus -a -um showy, handsome, semblance (*specere*, to look)
spectabilis -is -e admirable, spectacular, good-looking
specularius -a -um, speculum shining, mirror-like (speculum, a mirror)
speculatus -a -um shining, as if with mirrors
speir- twisted-
speiranthus -a -um with twisted flowers
speluncae, speluncarum (spelunchae) of caves, cave-dwelling
speluncatus -a -um, speluncosus -a -um cavitied, full of holes
Spergula Scatterer (l'Obel's name refers to the discharge of the seeds)
Spergularia Resembling-*Spergula*
sperm-, spermato-, -spermus -a -um seed- (σπερμα), -seed, -seeded
Spermatophtyta Seed-plants
sphacelatus -a -um necrotic, scorched, gangrened (σφακελος)
sphaer-, sphaero- globular-, spherical-, ball- (σφαιρα)
Sphaeralcea Spherical-*Alcea* (the shape of the fruit)
sphaerocephalon, sphaerocephalus -a -um round-headed
Sphagnum latinized from the Greek, σφαγνος, for a moss
sphegodes resembling wasps, σφηξ, σφηκος (flower shape)
spheno- wedge- (σφενος)
Sphenopteris Wedge-fern
sphondilius -a -um rounded
spica, spicati-, spicatus -a -um, spicifer -era -erum with an elongate inflorescence of sessile flowers, spiked, spicate (see Fig. 2(*a*))
spicant from the ancient German name, tufted (spikenard, spike, ear)
spica-venti ear of the wind, tuft of the wind
spiculi- spicule-, small-thorn-

Spigelia for Adrian van der Spiegel of Padua (1578–1625)
Spilanthes Stained-flower (receptacular marks of some species)
spilo- stained- (σπιλος)
spilofolius -a -um spotted-leaved (rose-hip fly = *Spilographa alternata*)
spina-christi Christ's thorn
spinescens, *spinifer -era -erum*, *spinifex*, *spinosus -a -um* spiny, with spines
spinulifer -era -erum, *spinulosus -a -um* with small spines
spir- twisted-, coiled- (σπειρα)
Spiraea Theophrastus' name, σπειραια, for a plant used for making garlands
spiralis -is -e, *spiratus -a -um* twisted, spiral
Spiranthes Twisted (the inflorescence)
spirellus -a -um small-coiled
spiro- twisted-, coiled-
splendens, *splendidus -a -um* gleaming, striking
spodo- ash-grey (σποδος)
spodochrus -a -um greyish-coloured, ashen
Spondias Theophrastus' name refers to the plum-like fruit
Spondianthus *Spondias*-flowered
spongiosus -a -um spongy
sponhemicus -a -um from Sponheim, Rhine
sporadicus -a -um scattered, widely dispersed (σπορας)
sporo-, *sporo-* spore-, seed-
Sporobolus Seed-caster
-sporus -a -um -seed, -seeded
sprengeri for Carl L. Sprenger (1846–1917), German nurseryman, Naples
spretus -a -um (sprettus) despised, spurned
spumarius -a -um foamy, frothing
spumescens becoming frothy, of frothy appearance
spumosus -a -um frothy
spurcatus -a -um fouled, nasty, filthy
spurius -a -um false, bastard, spurious

squalens, *squalidus -a -um* untidy, dingy, squalid

squamarius -a -um, *squamosus -a -um* scale-clad, covered with scales

squamatus -a -um with small scale-like leaves or bracts (*squamae*), squamate

squamigerus -era -erum scale-bearing

squarrosus -a -um rough (when leaves have protruding tips or sharp edges)

squillus -a -um shrimp-like, squill-like

stabilis -is -e firm, lasting, not changeable

stachy- spike-like-, resembling *Stachys*

Stachys Spike (the Greek name, σταχυς, used by Dioscorides for several dead-nettles)

-stachyon , *-stachys*, *stachyus -a -um* -spiked, -panicled

Stachytarpheta Thick-spike

Stachyurus Spiked-tail (the shape of the inflorescence)

stagnalis -is -e of pools

stagninus -a -um of swampy or boggy ground

stamineus -a -um with prominent stamens (stamen, a filament)

staminosus -a -um the stamens being a marked feature of the flowers

Stanhopea for Philip Henry, 4th Earl of Stanhope (1781–1855), President of the Medico-botanical Society

stans upright, erect, standing, self-supporting

Stapelia by Linnaeus for Johannes B. von Stapel, Dutch physician of Amsterdam

Staphylea Cluster, σταφυλη (a name in Pliny, refers to the bunched flowers)

-staphylos -bunch (as of grapes)

stasophilus -a -um living in stagnant water, loving stagnant waters

Statice Astringent (Dioscorides' name, στατικη, for the *Limonium* of gardeners)

Stauntonia for Sir George C. Staunton (1737–1801), Irish traveller in China

stauro- cross-shaped-, crosswise-, cruciform-

stegano-, *stego-* covered-over-, roofed- (στεγη)

steiro- barren- (στειρος)
Stellaria Star (the appearance of stitchwort flowers)
stellaris -is -e, stellatus -a -um star-like, with spreading rays, stellate
stelliger -era -erum star-bearing
stellipilus -a -um with stellate hairs
stellulatus -a -um small-starred, with small star-like flowers
-stemon -stamened (στημων)
sten-, steno- short-, narrow- (στενος)
Stenanthera Narrow-anthers
Stenocarpus Narrow-fruit (the flattened folicular fruits)
Stenochlaena Narrow-cloak (sporangia cover entire surface of the linear fertile pinnae)
Stenoloma Narrow-hem, λομα (the narrow indusium)
stenopetalus -a -um narrow-petalled
Stenotaphrum Narrow-trench, ταφρος (the florets are recessed into cavities in the rachis)
stephan-, stephano- crowned-, crown-, wreathed-, στεφανη
Stephanandra Male-crown (the arrangement of the persistent stamens)
Stephania for Frederick Stephan of Moscow (d. 1817)
Stephanotis Crowned-ear (the auricled staminal crown) also used by the Greeks for plants used for making chaplets or crowns
-stephanus -a -um -crowned
stepposus -a -um of the Steppes
Sterculia Dung (the evil-smelling flowers of some species)
stereo- solid-
sterilis -is -e infertile, barren, sterile
Sternbergia for Count Kaspar Moritz von Sternberg of Prague (1761–1838)
sternianus -a -um for Col. Sir Frederick Stern (1884–1967), horticultural pioneer of Highdown, Worthing [cultivarietal names 'Highdown' and *highdownensis*]
Stewartia for John Stewart, 3rd Earl of Bute (1713–1792), patron of botany

-stichus -a -um -ranked, -rowed (στιχος)
stict-, sticto- ,-stictus -a -um punctured-, -spotted (στικτος)
stictocarpus -a -um with spotted fruits
stigma- spot-, stigma- (στιγμα)
stigmaticus -a -um, -stigmus -a -um spotted, dotted, marked
stigmosus -a -um spotted, marked
Stilbe Shining
Stipa Tow (Greek use of the feathery inflorescences, like hemp, for caulking and plugging)
stipellatus -a -um with stipels (in addition to stipules)
stipitatus -a -um with a stipe or stalk
stipulaceus -a -um, stipularis -is -e, stipulatus -a -um, stipulosus -a -um with conspicuous stipules
stoechas Dioscorides' name for a lavender grown on the Iles d'Hyères, Toulon, which were called 'Stoichades'
Stokesia for Dr Jonathan Stokes (1755–1831), who worked with Withering on his arrangement of plants
stolonifer -era -erum spreading by stolons, with creeping stems which root at the nodes
-stomus -a -um -mouthed, στομα
Storax see *Styrax*
stragulatus -a -um, stragulus -a -um carpeting, mat-forming, covering
stramine-, stramineus -a -um straw-coloured
stramonii- *Stramonium*-like-
stramonium a name used by Theophrastus for the thorn apple, *Datura stramonium*
strangulatus -a -um constricting, strangling, with irregular constrictions
Stranvaesia for William Thomas Horner Fox-Strangways, Earl of Ilchester (1795–1865), botanist
strateumaticus -a -um forming an army, forming groups
Stratiotes Soldier (Dioscorides' name, στρατιωτης, for an Egyptian water plant with sword-shaped leaves)

Strelitzia for Queen Charlotte of Mecklenburg-Strelitz, wife of George III

strepens rustling, rattling

Strephonema Twisted-threads (the stamens)

strepsi-, *strept-*, *strepto-* twisted-, coiled-

Streptocarpus Twisted-fruit (στρεπτος), the fruits contort as they mature

Streptosolen Twisted-tube (the corolla tube is spirally twisted below the expanded part)

striatellus -a -um, *striatulus -a -um* somewhat marked with parallel lines, grooves or ridges

striatus -a -um ridged, striped, furrowed

stricti-, *stricto-*, *strictus -a -um* straight, erect, close, stiff, strict

Striga Swathe

strigatus -a -um straight, rigid, *Striga*-like

strigilosus -a -um with short rigid bristles (*strigilis*, a currycomb)

strigosus -a -um thin, lank, with rigid hairs or bristles, strigose

strigulosus -a -um somewhat strigose

striolatus -a -um faintly striped, finely lined

strobilaceus -a -um cone-like, cone-shaped (in fruit) (στροβιλος)

strobil-, *strobili-*, *strobilifer -era -erum* cone-bearing

strobus an ancient name for an incense-bearing tree (*Pinus strobus* has large seed-cones, στροβιλος)

Strobylanthes Cone-flower

Stromanthe Mattress-flower (στρωμα), the form of the inflorescence

strombuli- snail-shell-like-

strombuliferus -a -um bearing spirals, snail-like (as with the fruits of some *Medicago* species)

strongyl-, *strongylo-* round-, rounded-

Strophanthus Twisted-flower (the elongate lobes of the corolla)

strophio- turned-over, turning- (στροφο)

strumarius -a -um, *strumosus -a -um* cushion-like, swollen (signature for use in treatment of swollen necks)

strumi- cushion-like-swelling-, wen-, goitre-like-

strupi- strapped-

Struthiopteris Ostrich-feather (the fertile fronds)

Strychnos Linnaeus reapplied Theophrastus' name for poisonous solanaceous plants

stupeus -a -um, stuppeus -a -um woolly

stuposus -a -um, stuposus -a -um shaggy with matted tufts of long hairs, tousled, tow-like

stygia of the underworld, Stygean (*Globularia stygia* spreads by subterranean stolons), growing in foul water

stylo-, -stylus -a -um style-, -styled (στυλος)

stylosus -a -um with a prominent style

stypticus -a -um astringent, styptic (στυπτικος)

styracifluus -a -um flowing with gum

Styrax ancient Greek name for storax gum tree, *Styrax officinalis*

styrido- cruciform- (σταυρος)

Suaeda from the Arabic name

suaveolens sweet-scented

suavis -is -e sweet, agreeable

suavissimus -a -um very sweet-scented

sub-, suc-, suf-, sug- below-, under-, approaching-, nearly-, just-, less than-, usually-

subacaulis -is -e almost without a stem

subbiflorus -a -um resembling biflorus (a comparative relationship)

subcaeruleus -a -um slightly blue

subdiaphenus -a -um semi-transparent

suber corky (the ancient Latin name for the cork oak, *Quercus suber*)

suberectus -a -um growing at an angle, not quite upright

suberosus -a -um slightly bitten (sub-erosus), corky

sublustris -is -e glimmering, almost shining

submersus -a -um under-water, submerged

suboliferus -a -um bearing offspring (see *soboliferus*)

subsessilis -is -e very short stalked, almost-sessile

subterraneus -a -um below ground, underground

subtilis -is -e fine

Subularia awl (the leaf shape)

subulatus -a -um awl-shaped, needle-like, subulate

subuli- awl-shaped-

subulosus -a -um somewhat awl-shaped

succiferus -a -um producing sap, sappy

succiruber -era -erum with red sap

Succisa cut-off (the rhizome of *S. pratensis*)

succisus -a -um (*succisus*) cut off from below, abruptly ended

succosus -a -um full of sap, sappy

succotrinus -a -um from Socotra, Indian Ocean

Succowia for Georg Adolph Suckow (1751–1813) of Heidelberg

succubus -a -um lying upon (when a lower distichous leaf is overlain by the next uppermost leaf on the same side of the stem); Latin for a female nocturnal demon

succulentus -a -um fleshy, soft, juicy, succulent

sucidus -a -um, sucosus -a -um sappy, juicy

sudanensis -is -e from the Sudan, Sudanese

sudeticus -a -um from the Sudetenland of Czechoslovakia and Poland

suecicus -a -um from Sweden (*Swabia*), Swedish

Suaeda from the Arabic, suwed-mullah, for *Sueda baccata*

suffocatus -a -um suffocating (the flower heads of *Trifolium suffocatum* turn to the ground)

suffruticosus -a -um somewhat shrubby at the base, soft-wooded and growing yearly from ground level

suffultus -a -um supported, propped-up

suionum of the Swedes (*Sviones*)

sulcatus -a -um furrowed, grooved, sulcate

sulfureus -a -um, sulphureus -a -um pale-yellow, sulphur-yellow

sultani for the Sultan of Zanzibar

sumatranus -a -um from Sumatra, Indonesia

super-, supra- above-, over-

superbus -a -um magnificent, proud, superb

superciliaris -is -e eyebrow-like, with eyebrows, with hairs above *superfluus -a -um* overflowing

supernatans living on the surface of water

supinus -a -um lying flat, extended, supine

supra- above-, on-the-surface-of-

suprafolius -a -um growing on a leaf

supranubius -a -um of very high mountains, from above the clouds

surattensis -is -e from Surat on the west coast of India

surculosus -a -um shooting, suckering, freely producing young shoots

surrectus -a -um not quite upright or erect, leaning (sub-rectus)

sursum- forwards-and-upwards-

susianus -a -um from Susa, Iran

suspendus -a -um, *suspensus -a -um* lax, hanging down, pendent, suspended

sutchuensis -is -e from Szechwan, China

Sutherlandia for James Sutherland (d. 1719), Superintendant of Edinburgh Botanic Garden and botanical writer

sy-, *syl-*, *sym-*, *syn-*, *syr-*, *sys-* with-, together with-, united-, joined-

sycamorus fig-fruited, of the fig (*Ficus sycamorus*)

sychno- many-times-, frequent- (συχνος)

syco-, *sycon-* fig-like-fruit-, fig- (συκον)

Sycopsis Fig-resembler (συκον) looks like some shrubby *Ficus*

sylvaticus -a -um, *sylvester -tris -tre* wild, of woods or forests, sylvan

sylvicolus -a -um inhabiting woods

sympho-, *symphy-* growing-together- (συμφυτος)

Symphonia Grown-together-stamens (they are united with five groups of three linear anthers alternating with the stigmatic lobes)

Symphoricarpos (us) Clustered-fruits, συμφερω–καρπος (the conspicuous berries)

Symphytum Grow-together-plant (Dioscorides' name for healing plants, including comfrey, *conferva* of Pliny)

Symplocos United (the united stamens)

syn- together-

synciccus -a -um with flowers of different sexes in the same inflorescence

syphiliticus -a -um of syphilis, after a character in a 16th century Latin poem on pox (*Lobelia syphilitica* used to treat the disease)

syriacus -a -um from Syria, Syrian

Syringa Pipe (formerly for *Philadelphus* but re-applied by Dodoens, use of the hollow stems to make flutes)

syringanthus -a -um lilac-flowered, *Syringa*-flowered

syzigachne with scissor-like glumes

Syzygium Paired (from the form of branching and opposite leaves. Formerly applied to *Calyptranthus*)

tabacicomus -a -um with a tobacco-coloured head

tabacinus -a -um tobacco-like

tabacum (tabaccum) from the Mexican vernacular name for the pipe used for smoking the leaves of *Solanum*

Tabebuia from a Brazilian vernacular name

tabernaemontanus -a -um for J.T. Bergzabern of Heidelberg (d. 1590), physician and herbalist (his latinization of Bergzabern)

tabescens wasting-away

tabulaeformis -is -e, tabuliformis -is -e flat and circular, plate-like

tabularis -is -e, tabuli- table-flat, flattened

tabulatus -a -um layer upon layer

tacamahacca from an Aztec vernacular name for the resin from *Populus tacamahaca*

tacazzeanus -a -um from the Takazze River, Ethiopia

Tacca from a Malayan name, taka, for arrowroot

Taccarum *Tacca-arum* (implies intermediate looks but not hybridity)

taccifolius -a -um with leaves like *Tacca*

tactilis -is -e sensitive to touch

taeda an ancient name for resinous pine trees

taediger -era -erum torch-bearing

taediosus -a -um loathsome

taegetus -a -um from Mt Taygetos, Greece

taenianus -a -um shaped (segmented) like a tapeworm, *Taenia*-like
Taeniopteris Ribbon-fern
taeniosus -a -um ribbon-like, banded (leaves)
Tagetes for Tages, Etruscan god and grandson of Jupiter
tagliabuana for the brothers Tagliabe
taipeicus -a -um from Taipei Shan, Shensi, China
taiwanensis -is -e from Formosa (Taiwan), Formosan
tamarici-, tamarisci- *Tamarix*-like-
tamarindi- tamarind-like
Tamarindus Indian-date (from the Arabic, tamr)
Tamarix the ancient Latin name, for the Spanish area of the River Tambo (*Tamaris*)
tamnifolius -a -um bryony-leaved, with leaves like *Tamus* (*Tamnus* of Pliny)
Tamus from the name in Pliny for a kind of vine
tanacet-, tanaciti- tansy-like-, *Tanacetum-*
Tanacetum Immortality (tansy was placed amongst the winding sheets of the dead to repel vermin)
tanaiticus -a -um from the region of the River Tanais (Don) in Sarmatia
Tanakaea for Yoshio Tanaka, Japanese botanist
tanguticus -a -um of the Tangut tribe of Gansu, NE Tibet, Tibetan
tapein-, tapeino- humble-, modest-
Tapeinanthus Low-flower; refers to the small stature
Tapeinochilus Modest-lip; refers to the small labellum
tapesi- carpet-, ταπης
tapeti- carpet-like-
taphro- ditch- (ταφρος)
Tapiscia an anagram of *Pistacia*
Tapura from the vernacular name in Guiana
taraxaci- dandelion-like-, *Taraxacum-*
Taraxacum Disturber (from the Persian name for a bitter herb)
tardi-, tardus -a -um slow, reluctant, late
tardiflorus -a -um late-flowering

tardivus -a -um   slow-growing

tartareus -a -um, tartrus -a -um   of the underworld, with a loose crumbling surface

tartaricus -a -um, tataricus -a -um   from Tartary, Central Asia

tasmanicus -a -um   from Tasmania

tatsiensis -is -e   from Tatsienlu, China

tatula   from an old name for a *Datura*

tauricus -a -um   from the Crimea (*Tauria*)

taurinus -a -um   from Turin, Italy, or of bulls

tax-, taxi-, taxo-   orderly-, order- (ταξις)

taxi-   yew-like-, resembling *Taxus*

taxodioides   resembling *Taxodium*

Taxodium   Yew-like, resembling *Taxus*

taxoides   resembling yew

taxonomy   orderly law, classification

Taxus   the ancient Latin name for the yew

taygeteus -a -um   from Mt Taygetos, Greece

tazetta   little cup (the corona of *Narcissus tazetta*)

technicus -a -um   special, technical

Tecoma, Tecomaria   from the Mexican name of the former

Tectaria   Roofed (the complete indusium)

Tectona   from the Tamil name, tekka, for teak

tectorum   of rooftops, growing on rooftops, of the tiles

tectus -a -um   with a thin covering, hidden, tectate

Teesdalia   for Robert Teesdale, Yorkshire botanist

tef   the Arabic name for *Eragrostis abyssinica* (tef grass)

tegens, tegetus -a -um   mat-like

tegumentus -a -um   covered (e.g. indusiate)

tel-, tele-   far-, far-off-, afar- (τηλε)

telephioides   resembling *Sedum telephium*

Telephium   Distant-lover. A Greek name,τηλεφιον, for a plant thought to be capable of indicating reciprocated love

teleuto-, telio-   terminal-, completion-, an end- (τελευτη)

Tellima   an anagram of *Mitella*

telmataia, *telmateius -a -um* of marshes, of muddy water (τελμα, τελματος)
telonensis -is -e from Toulon (*Telenium*), France
Telopea Seen-at-a-distance (τελοπας), the conspicuous crimson flowers
temenius -a -um of sacred precincts or holy places
temulentus -a -um drunken, intoxicating (toxic seed of ryegrass)
temulus -a -um synonymous with temulentus (the rich fragrance of rough chervil)
tenax gripping, stubborn, firm, persistent, tenacious
tenebrosus -a -um somewhat tender, of shade, dark, gloomy
tenens enduring, persisting
tenellus -a -um delicate, tenderish
teneri-, *tenerus -era -erum* soft, tender, delicate
tentaculosus -a -um with sensitive glandular hairs
tenui-, *tenuis -is -e* slender, thin, fine
tenuifolius -a -um slender-leaved
tenuior more slender
tephro-, *tephrus -a -um* ash-grey-, ashen (τεφρα)
Tephrosia Ashen (the leaf colour)
ter- three-times, triple-, thrice-
terato- prodigious-, monstrous- (τερας)
terebinthi- *Pistacia*-like-, turpentine-
terebinthifolius -a -um with leaves like those of *Pistacia terebinthus*
terebinthinus a former name for Chian turpentine tree, *Pistacia terebinthus*
teres -etis -ete, *tereti-* quill-like, cylindrical, terete
tereticornis -is -e with cylindrical horns
teretiusculus -a -um somewhat smoothly rounded
tergeminus -a -um three-paired
tergi- at the back-
Terminalia Terminal (the leaves are frequently crowded at the ends of the branches)
terminalis -is -e terminal (the flower on the stem)

ternateus -a -um from the Ternate Islands, Moluccas

ternatus -a -um, ternati-, terni- with parts in threes, ternate (see Fig. 5(*e*))

Ternstroemia for Christopher Ternstroem (d. 1754), Swedish naturalist in China

terrestris -is -e growing on the ground, not epiphytic or aquatic

tersi- neat-

tertio- third-

tesquicolus -a -um of waste land, of desert land

tessellatus -a -um (tesselatus) chequered, mosaic-like, tessellated

testaceus -a -um brownish-yellow, terracotta, brick-coloured

testicularis -is -e, testiculatus -a -um tubercled, having some testicle-shaped structure (e.g. a tuber or fruit)

testudinarium resembling tortoise shells

teter -era -erum having a foul smell

tetra- square-, four- (τετρα)

Tetracentron Four-spurs (the spur-like appendages of the fruit)

Tetracme Four-points (the shape of the fruit)

Tetradymia Fourfold (the groups of flowers and their involucral bracts)

Tetragonolobus Quadrangular-pod (the fruit)

tetragonus -a -um four-angled, square

tetrahit four-times (tetraploid), foetid

tetralix a name, τετραλιξ, used by Theophrastus for the cross-leaved state when the leaves are arranged in whorls of four

tetrandrus -a -um with four stamens, four-anthered

tetraplus -a -um fourfold (e.g. ranks of leaves)

tetraquetrus -a -um sharply four-angled (*Arenaria tetrequetra*)

Tetraspis Four-shields

Tetrastigma Four-stigmas (the four-lobed stigma)

Teucrium Dioscorides' name, τευκριον, perhaps for Teucer, hero and first King of Troy

texanus -a -um, texensis -is -e from Texas, USA, Texan

textilis -is -e used for weaving

thalami- bedchamber-, θαλαμος, receptacle-
thalassicus -a -um, thalassinus -a -um sea-green, growing in the sea
thalidi-, thallo thallus- (a vegetative body without differentiation into stem and leaves)
Thalia, thalianus -a -um for Johann Thal, German botanist (1542–1583). Thalia was also one of the three Graces
Thalictrum a name, θαλικτρον, used by Dioscorides for another plant
thamn-, thamno-, -thamnus -a -um -shrub-like, -shrubby (θαμνος)
Thapsia ancient Greek name, θαψος, used by Theophrastus
thapsus from the Island of Thapsos (old generic name, θαψος, for *Cotinus coggygria*)
Thea the latinized Chinese name, T'e
thebaicus -a -um from Thebes, Greece
-theca, theco-, -thecus -a -um box-, -chambered, -cased (θηκη)
Thecacoris Split-cells (the anthers)
-thecius -a -um -cased, θηκη, -chambered
theifer -era -erum tea-bearing
theio- smoky-
theioglossus -a -um smoke-tongued
thele-, thelo-, thely- female-, nipple- (θηλυς)
thelephorus -a -um covered in nipple-like prominences
Thelycrania the name, θηλυκρανεια, used by Theophrastus
Thelygonum Girl-begetter (claimed by Pliny to cause girl offspring)
Thelypteris (Thelipteris) Female-fern (θηλυς–πτερις) Theophrastus' name for a fern
Themeda from an Arabic name
Theobroma God's-food
theoides resembling tea-plant, *Thea*-like
theriacus -a -um (theriophonus, for theriophobus?) antidote (theriacs are antidotes to poisons and bites of wild beasts, θηρ)
thermalis -is -e of warm springs (θερμη)
Thermopsis Lupin-like (θερμος)
thero- summer- (θερος)

Thesium a name, θησειον, in Pliny for a bulbous plant
Thespesia Divine (commonly cultured round temples)
thessalonicus -a -um, thessalus -a -um from Thessaly
Thevetia for Andre Thevet (1502–1592), French traveller in Brazil and Guiana
thibeticus -a -um from Tibet
thigmo- touch-, θιγγανω
thino- sand-
thirsi- panicled-
Thladiantha Eunuch-flower (female flowers have aborted stamens)
Thlaspi the name, θλασπις, used by Dioscorides for cress
thora of corruption, of ruination (a medieval name for a poisonous buttercup)
Thrinax, -thrinax Fan, -fanned, -trident
-thrix -hair, -haired (θριξ, τριχος)
Thuidium *Thuja*-like
Thuja, Thuya Theophrastus' name, θυια, for a resinous fragrant-wooded tree
Thujopsis, Thuyopsis Resembling-*Thuja*
Thunbergia for Karl Pehr Thunberg (1743–1822), of Uppsala
thurifer -era -erum, turifer -era -erum incense-bearing, frankinsense-producing
thuringiacus -a -um from mid-Germany (*Thuringia*)
thuyioides, thyoides *Thuja*-like
thymbra an ancient Greek name for a savory thyme-like plant
Thymelaea Thyme-olive (the leaves and fruit)
Thymus Theophrastus' name, θυμος, for a plant used in sacrifices
thyrs-, thyrsi- contracted-panicle-, θυρσος baccic staff
thyrsoideus -a -um with a pyramidal panicle-, thyrsoid (see Fig. 3(*d*))
thysano-, thysanoto- fringed-
Tiarella Little-diadem (τιαρα), the capsules
tibeticus -a -um from Tibet
tibicinis piper's or flute-player's
tibicinus -a -um hollow-reed-like, flute-like

Tibouchina from a Guianese name
Tigridia Tiger (the markings of the perianth)
tigrinus -a -um striped, spotted, tiger-toothed
tigurinus -a -um from Zurich (*Turicum*)
Tilia Wing (the ancient Latin name for the lime tree)
tiliae-, tiliaceus -a -um lime-like, resembling *Tilia*
tillaea from a former generic name, *Tillaea*, for a *Crassula*
Tillandsia for Elias Tillands, Swedish botanist and Professor of Medicine
tinctorius -a -um used for dyeing
tinctorum of the dyers
tinctus -a -um coloured
tingens stained, dyed, staining
tingitanus -a -um from Tangiers, Morocco
tini- *Tinus*-like (*Viburnum*-like)
tinus the old Latin name for *Laurustinus* (*Viburnum*)
tipuliformis -is -e resembling a *Tipulid* (cranefly)
tirolensis -is -e from the Tyrol, Tyrolean
titano- chalk-, lime-
titanum, titanus -a -um of the Titans, gigantic, very large
Tithonia after Tithonus from Greek mythology, brother of Priam
tithymaloides spurge-like
Tithymalus an ancient name, τιθυμαλλος, used in Pliny for plants with latex, spurges (*Tithymallus*)
-tmemus -a -um -free
toco- offspring-
Todea for H.J. Tode (1733–1797), German mycologist
tofaceus -a -um tufa-coloured, gritty
Tofieldia for Thomas Tofield (1730–1779), Yorkshire naturalist
togatus -a -um robed, gowned
Tolmiea for Dr William F. Tolmie of the Hudson's Bay Company
Tolpis a name of uncertain derivation
toluiferus -a -um producing balsam of tolu (*Myroxylon toluifera*)
tomentellus -a -um somewhat hairy

tomentosus -a -um thickly matted with hairs
tomi-, -tomus -a -um cutting-, -cut, -incised
tonsus -a -um shaven, sheared, shorn
tophaceus -a -um see *tofaceus*
Tordylium the name, τορδυλιον, used by Dioscorides
Torenia for Rev. Olof Torén (1718–1753), chaplain to the Swedish East India Company in India, Surat and China
torfosus -a -um growing in bogs
Torilis a meaningless name by Adanson
toringoides toringo-like (Japanese name for a *Malus*)
torminalis -is -e of colic (used medicinally to relieve colic)
torminosus -a -um causing colic
torosus -a -um cylindrical with regular constrictions
torquatus -a -um with a (chain-like) collar, necklaced
Torreya for Dr John Torrey (1796–1873), American botanist, contributor to the *Flora of North America*
torridus -a -um of very hot places
torti-, tortilis -is -e, tortus -a -um twisted
tortuosus -a -um meandering (irregularly twisted stems)
Tortula Twisted (the 32 spirally twisted teeth of the peristome)
torulosus -a -um swollen or thickened at intervals
torvus -a -um fierce, harsh, sharp
tovarensis -is -e from Tovar, Venezuela
Townsendia for David Townsend (*c.* 1840), Pennsylvanian botanist, USA
toxi-, toxicarius -a -um, toxicus -a -um toxic, containing a poisonous principle (*toxicum*)
toxifera -a -um poison-bearing, poisonous
Toxicodendron (um) Poison-tree
toza from a South African native name
trabeculatus -a -um cross-barred
trachelium neck (old name for a plant used for throat infections)
trachelo- neck-
Trachelospermum Necked-seed

trachy- shaggy-, rough- (τραχυς)
Trachycarpus Rough-fruit
trachyodon short-toothed, rough-toothed
Trachystemon Rough-stamen
Tradescantia for John Tradescant (d. 1638) (son of Old John Tradescant), gardener to Charles I
tragacantha yielding gum-tragacanth (from a Greek plant name – goat-thorn, *Astragalus tragacantha*)
trago- goat- (τραγος)
tragoctanus -a -um goat's-bane
Tragopogon Goat-beard (the pappus of the fruit)
Tragus Goat
trajectilis -is -e, *trajectus -a -um* passing over (separation of anther loculi by connective)
trans- through-, beyond-, across-
transalpinus -a -um crossing the Alps
transiens intermediate, passing-over
translucens almost transparent
transversus -a -um athwart, across, collateral
transwallianus -a -um from Pembroke, South Wales (beyond Wales)
transylvanicus -a -um from Romania (Transylvania)
Trapa from *calcitrapa*, a four-spiked weapon used in battle to maim cavalry horses' hooves
trapesioides lozenge-shaped, shaped like a deformed square, tapeziod, τραπεζιον
Tremella Quiverer
tremuloides aspen-like, resembling *Populus tremula*
tremulus -a -um trembling, shaking, quivering
trepidus -a -um restless, trembling
tri- three- (τρεις)
triacanthos, *triacanthus -a -um* three-spined
triandrus -a -um three-stamened
triangulari-, *triangularis -is -e*, *triangulatus -a -um* three-angled, triangular

tricamarus -a -um three-chambered
trich-, tricho-, -trichus -a -um hair-like-, -hairy (θριξ, τριχος)
trichoides hair-like
Trichomanes Hair-madness (τριχο-μανες) Theophrastus' name for maidenhair spleenwort
Trichophorum Hair-carrier (perianth bristles)
Trichosanthes Hair-flower (the fringed corolla lobes)
trichospermus -a -um hairy-seeded
trichotomus -a -um three-forked, triple-branched
tricoccus -a -um three-seeded, three-berried
tricolor three-coloured
tricornis -is -e, tricornutus -a -um with three horns
Tricyrtis Three-domes (τρεις–κυρτος), the form of the bases of the three outer petals
tridactylites three-fingered, τρεισδακτυλον
Tridax Three-toothed (Theophrastus' name, θριδαξ, for a lettuce, ligulate florets are 3-fid)
triduus -a -um lasting for three days
triennialis -is -e, triennis -is -e lasting for three years
trientalis -is -e a third of a foot in length, about four inches tall
trifasciatus -a -um three-banded
trifidus -a -um divided into three, three-cleft
Trifolium Trefoil (the name in Pliny for trifoliate plants)
trifurcatus -a -um three-forked
triglans three-nutted-fruits, containing three nuts
Triglochin Three-barbed, γλωχις (the fruits)
Trigonella Triangle (the flower of fenugreek seen from the front)
trigonus -a -um three-angled, with three flat faces and angles between them
Trillium In-threes (the parts are conspicuously in threes, lily-like)
trimestris of three months, maturing in three months (*Lavatera trimestris*)
trimus -a -um lasting three years

trinervius -a -um three-nerved (three-veined leaves)
trineus -a -um, *trinus -a -um* in threes
Trinia for K.B. Trinius (1778–1844), Russian botanist
trionus -a -um three-coloured
tripartitus -a -um divided into three segments
Tripetaleia Three-petals (the tripartite floral arrangement)
triphyllos three-leaved, with three leaflets
Tripleurospermum Three-ribbed-seed
tripli-, *triplo-* triple-, threefold-, *triplus*
Triplochiton Three-coverings (the flowers have a series of petaloid staminodes within the staminal ring, forming the third layer)
Tripterygium Three-wings (the three-winged fruits)
triqueter, *triquetrus -a -um* three-cornered, three-edged, three-angled (stems)
Trisetum Three-awns
tristis -is -e bitter, sad, gloomy, dull-coloured
trisulcus -a -um three-grooved
Triteleia Triplicate (the flower parts are in threes)
triternatus -a -um three times in threes (division of the leaves)
Triticum the classical name for wheat
tritifolius -a -um with polished leaves
Tritonia Weathercock (τριτον) the disposition of the stamens
tritus -a -um in common use
triumphans exultant, triumphal
triumvirati of three men (like mayoral regalia)
trivialis -is -e common, ordinary, wayside, of crossroads
trixago *Trixis*-like
Trixis Triple (three-angled fruits)
Trochetia for R.I.G. du Trochet (1771–1847), French plant physiologist
Trochetiopsis *Trochetia*-like
trocho- wheel-like-, hooped-, wheel- (τροχος)
Trochodendron Wheel-tree (the radiately spreading stamens)

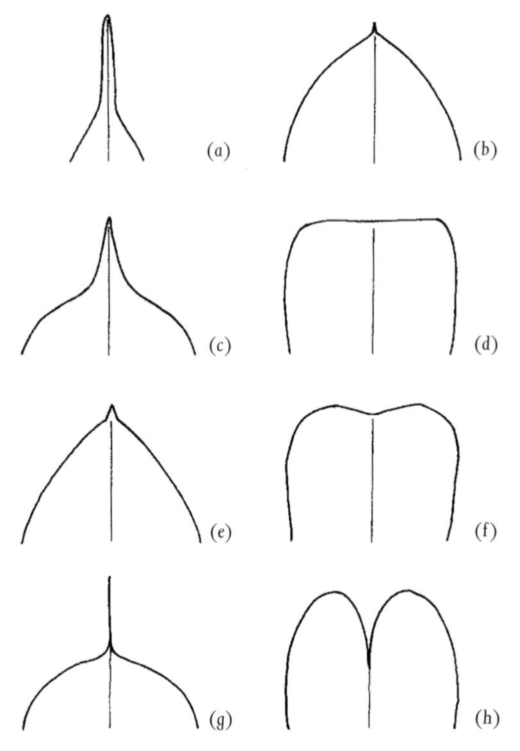

Fig. 7. Leaf-apex shapes which provide specific epithets.
(*a*) Caudate (e.g. *Ornithogalum caudatum* Jacq.) with a tail;
(*b*) mucronate (e.g. *Erigeron mucronatus* DC.) with a hard tooth;
(*c*) acuminate (e.g. *Magnolia acuminata* L.) pointed abruptly;
(*d*) truncate (e.g. *Zygocactus truncatus* K.Schum.) bluntly foreshortened;
(*e*) apiculate (e.g. *Braunsia apiculata* Schw.) with a short broad point;
(*f*) retuse (e.g. *Daphne retusa* Hemsl.) shallowly indented;
(*g*) aristate (e.g. *Berberis aristata* DC.) with a hair-like tip, not always restricted to describing the leaf-apex;
(*h*) emarginate (e.g. *Limonium emarginatum* (Willd.) O.Kuntze) with a deep mid-line indentation.

troglodites wren, of caves

troglodytarum of cave-dwellers, apes or monkeys (cf. *sapientium*. Implies inferiority or unsuitability for man)

Trollius Closed-in-flower (*flos trollius*, from the Swiss-German name, trollblume, for the globe flower)

Tropaeolum Trophy (τροπαιον) the gardener's *Nasturtium* was likened by Linnaeus to the losers' shields and helmets displayed by the Greeks after victories in battle

-trophus -a -um -nourished (τροφις)

-tropis -keeled (τροπις)

-tropus -a -um -turning (τροπη)

tropicus -a -um of the tropics, tropical

trulliferus -a -um with leaves shaped like a scoop or trowel

trullis -is -e ladle-shaped, scoop-shaped

truncatulus -a -um, *truncatus -a -um* blunt-ended (the apex of a leaf (see Fig. 7(*d*)), truncate

truncicolus -a -um tree-trunk-dweller

tsintauensis -is -e from Tsingtau, Kwangsi Chuan, China

Tsuga from the Japanese vernacular name for the hemlock cedar

tsu-shimense from Tsu-shima, Japan

Tsusiophyllum *Tsusia*-leaved (from the Japanese, tsutsutsi, a subgeneric group of *Rhododendron*)

tubaeflorus -a -um with trumpet-shaped flowers

tubatus -a -um trumpet-shaped

Tuberaria Tuber (rootstock of *Tuberaria vulgaris*)

tuberculatus -a -um, *tuberculosus -a -um* knobbly, warted, warty, tuberculate (the surface texture)

tubergenianus -a -um, *tubergenii* for the van Tubergen bulb-growers of Holland

tuberosus -a -um swollen, tuberous

tuberiferus -a -um bearing tubers

tubi- tube-, pipe-

tubifer -era -erum, *tubulosus -a -um* tubular, bearing tubular structures

tubiflorus -a -um with trumpet-shaped flowers
tubiformis -is -e tube-shaped, tubular
tubulosus -a -um large-tubular
tucumaniensis -is -e from Argentina, Argentinian
tuitans guarding (of leaves that adopt a sleep-position)
tul- warted-
Tulbaghia for Rijk Tulbagh (1699–1771), one time Governor of the Cape of Good Hope
Tulipa from the Persian name, dulband, for a turban
tulipi- tulip-, *Tulipa*-like-
tulipiferus -a -um tulip-bearing, having tulip-like flowers
tumescens inflated, puffed-out, tumescent
tumidi-, tumidus -a -um swollen, tumid
tumulorum of burial mounds, of tumuli
tunbrigensis -is -e from Tonbridge Wells
Tunica Undergarment (the bracts below the calyx)
tunicatus -a -um coated, having a covering or tunic, tunicate
tuolumnensis -is -e from Tuolumne river and county, California, USA
tupi-, tupis- mallet-like-
turbinatus -a -um, turbiniformis -is -e top-shaped, turbinate
turcicus -a -um from Turkey, Turkish
turcomanicus -a -um from Turkestan
turcumaniensis -is -e from Turku, Finland
turfaceus -a -um, turfosus -a -um growing in bogs
turgescens becoming distended, becoming turgid
turgidus -a -um inflated, turgid
turgiphalliformis -is -e erect-phallus-shaped
turio- sucker-, scaly-shoot-
Turnera for William Turner (1508–1568), Tudor botanist of Wells, author of *A New Herbal*
turpis -is -e ugly, deformed
Turraea for Georgio della Turra (1607–1688), Professor of Botany at Padua
Turrita, Turritis Tower

turritus -a -um   turreted

Tussilago   Coughwort (medicinal use of leaves for treatment of coughs)

tycho-   by chance-

tylicolor   dark-grey, coloured like a woodlouse

tylo-   knob-, callus-, swelling-

tympani-   drum-

Typha   a Greek name, τυφη, used by Theophrastus for various plants

typhinus -a -um, typhoides   bulrush-like, resembling *Typha*, relating to fever

typicus -a -um   the type, typical

tyrianthinus -a -um   purple-coloured (Tyrian purple)

Uapaca   from the Madagascar vernacular name

uber -is -e   luxuriant, full, fruitful

ucranicus -a -um   from the Ukraine, Ukrainian

udensis -is -e   from the River Uda or the Uden district of Siberia

uduensis -is -e   from Udu, New Guinea

ugandae, ugandensis -is -e   from Uganda, East Africa

ugni   from a Chilean name for *Ugni molinae* (*Myrtus ugni*)

-ugo   -having (a feminine suffix in generic names)

ugoensis -is -e   from Mt Ugo, Luzon, Philippines

ulcerosus -a -um   knotty, lumpy

-ulentus -a -um   -abundant, -full

Ulex   an ancient Latin name in Pliny

ulicinus -a -um, ulicoides   resembling *Ulex*

uliginosus -a -um   marshy, of swamps or marshes

-ullus -a -um   -smaller, -lesser

Ulmaria   Elm-like (Gesner's name refers to the appearance of the leaves)

ulmi-, ulmoides   elm-like, resembling *Ulmus*

Ulmus   the ancient Latin name for elms

ulo-   shaggy- (ουλος)

-ulosus -a -um   minutely-, somewhat-

Ulothrix Shaggy-hair (the coarse filaments of this green alga)
ultonius -a -um from Ulster
ultra- beyond-, more than-
-ulus -a -um -tending to, -having somewhat
ulvaceus -a -um resembling the green alga *Ulva*, sea-lettuce
umbellatus -a -um with the branches of the inflorescence all rising from the same point, umbellate (see Fig. 2(*e*))
umbelli- umbel- (*umbella*, a sunshade)
Umbellularia Little-umbel (the inflorescences)
umbilicatus -a -um, umbilicus -a -um navelled, with a navel
Umbilicus Navel (e.g. the depression in the leaf surface above the insertion of the petiole)
umbo- knob-like-
umbonatus -a -um with a raised central boss or knob
umbracul- umbrella-like-
umbraculiferus -a -um shade-giving, umbrella-bearing (e.g. large leaves)
umbrosus -a -um growing in shade, shade-loving
umidus -a -um moist
un- one-, single-, not-
Uncaria Hooked, crooked, with hooked tips (hooked leaf tips)
uncatus -a -um, uncus with hooks, hook-like, hooked (*uncatus*)
unci- hook-, *uncatus* hooked
Uncifera hook-bearer
uncinatus -a -um with many hooks, barbed
Uncinia Much-hooked (the bur-like fruiting heads)
unctuosus -a -um with a smooth shiny surface, greasy
undatus -a -um, undosus -a -um not flat, billowy, waved, undulate
undulatus -a -um wavy
unedo the Latin name for the *Arbutus* tree and its fruit meaning 'I eat one'
ungui- clawed-
unguicularis -is -e, unguiculatus -a -um with a small claw or stalk (e.g. the petals)

ungulatus -a -um clawed
unguilobus -a -um with claw-like lobes (the leaf margins)
uni-, *unio-* one-, single-
uniflorus -a -um one-flowered
unilateralis -is -e one-sided, unilateral
unilocularis -is -e with a one-chambered ovary
unioloides resembling *Uniola* (American sea oats)
unitus -a -um joined, united
uplandicus -a -um from Uppland, Sweden
uporo Fijian vernacular name of cannibal tomato (*Solanum anthropophagorum*)
uragogus -a -um diuretic
uralensis -is -e from the Ural Mountains, Russia
uralum from vernacular name, urala swa, for *Hypericum uralum*
uranthus -a -um with tailed flowers
urbanii for Ignaz Urban (1848–1931), of Berlin Botanical Museum
urbanus -a -um, *urbicus -a -um* of the town, urban
urceolaris -is -e, *urceolatus -a -um* pitcher-shaped, urn-shaped
Urciolena Small-pitcher (the flower shape)
Uredo Blight (from the scorched appearance of infected host plants)
Urena from the Malabar name
urens acrid, stinging, burning, *uro* to burn
Urera Burning (cow itch)
Urginea from the Algerian type locality, the area of the tribe Beni Urgin
urnigerus -a -um urn-bearing
uro-, *-urus -a -um* tail-, -tailed (ουρα)
urophyllus -a -um tail-leaved
urseolatus -a -um crowded, *urgeo* to throng
Ursinia for Johannes Ursinus (1608–1666), author of *Arboretum Biblicum*
ursinus -a -um bear-like (the smell), northern (under the *Ursa major* constellation)
urtic-, *urticae-* nettle-, *Urtica-*

Urtica Sting (the Latin name)
-urus -a -um -tailed
Urvillea see *Durvillaea*
-usculus -a -um -ish (a diminutive ending)
usitatissimus -a -um most useful
usitatus -a -um everyday, ordinary, useful
Usnea a name of uncertain meaning by Adanson
usneoides resembling *Usnea*, hanging in long threads
ustulatus -a -um scorched-looking
utilis -is -e useful
Utricularia Little-bottle (*utriculus*), the underwater traps of the bladderwort
utricularis -is -e, utriculatus -a -um with utricles, bladder-like
utriculosus -a -um bladder-like, inflated
utriger -era -erum bearing bladders
-utus -a -um -having
uva-crispa curly-bunch (derivation doubtful)
uvaria from an old generic name, clustered (*uva*, a bunch of grapes)
Uvariodendron *Uvaria*-like-tree
Uvariopsis Like-*Uvaria*
uva-ursi bear's-berry (Latin equivalent of the name *Arctostaphylos*)
uva-vulpis fox's-berry
uvidus -a -um damp, moist
uvifer -era -erum fruiting in clusters, grape-bearing
uviformis -is -e in a clustered mass, like a cluster of swarming bees
Uvularia Palate (*uvula*), either from the hanging flowers or the fruits

Vaccaria Cow-fodder (an old generic name from *vacca*, a cow)
vaccini-, vaccinii-, vaccinioides bilberry-like, resembling *Vaccinium*
vaccinifolius -a -um with *Vaccinium*-like leaves
vacciniiflorus -a -um with *Vaccinium*-like flowers
Vaccinium a Latin name of great antiquity (like *Hyacinthus*) with no clear meaning
vaccinus -a -um the colour of a red cow, of cows

vacillans variable, swinging, versatile
vagans of wide distribution, wandering
vagensis -is -e from the River Wye (*Vaga*)
vaginans, *vaginatus -a -um* having a sheath, sheathed (as the stems of grasses by the leaf-sheaths)
vaginervis -is -e with veins arranged in no apparent order
vagus -a -um uncertain, wandering, varying, inconstant
vaillantii, *valantia* for S. Vaillant (Valantius) (1669–1722), French botanist
valdivianus -a -um, *valdiviensis -is -e* from Valdivia, Chile
valentinus -a -um from Valencia, Spain
Valeriana Health (*valere*), from a medieval name for its medicinal use
Valerianella diminutive of the name *Valeriana*
valesiacus -a -um see *vallesiacus*
validus -a -um, *validi-* well-developed, strong
valerandi for Dourez Valerand, 16th century botanist
vallerandii for Eugene Vallerand
vallesiacus -a -um, *vallesianus -a -um* from Valais (*Wallis*), Switzerland
vallicolus -a -um living in valleys
Vallisneria for Antonio Vallisneri de Vallisnera (1661–1730), Professor at Padua
Vallota for Pierre Vallot (1594–1671), French botanist and garden writer
valverdensis -is -e from Valverde, Hierro, Canary Isles
valvulatus -a -um articulated, jointed
Vancouveria for Capt. George Vancouver (1757–1798), on the *Discovery's* exploration of NW coastal America
Vanda from the Sanskrit name
vandasii for Dr Carl Vandas (1861–1923), Professor of Botany at Brno
Vangueria from the Madagascan vernacular name,voa-vanguer
Vanieria for J. de Vanier (1664–1739), French Jesuit and author of *Praedum Rusticum*
Vanilla Little-sheath (from the Spanish name describing the fruit)

vari-, *varii-* differing, changing, diverse, varying
variabilis -is -e, *varians* variable, not constant
variatus -a -um several, various
varicosus -a -um with dilated veins or filaments, varicose
variegatus -a -um irregularly coloured, blotched, variegated
variifolius -a -um variable-leaved
variolaris -is -e, *variolatus -a -um* pock-marked, pitted
variolosus -a -um very pock-marked, with large dimples
varius -a -um variable, changing, variegated
vas-, *vasi-* duct- (tube), vessel- (container)
vasconicus -a -um from the Basque country
vascularis -is -e possessing vessels (of the conductive tissue)
vasculosus -a -um shaped like a large dish or pitcher
vasculum a small vessel (container used by field botanists)
vastus -a -um vary large, vast
vectensis -is -e from the Isle of Wight (*Vectis*)
vegetus -a -um growing strongly or quickly, vigorous
Veitchia, *veitchii* for John Veitch (1725–1839) and his successors of Exeter and Chelsea
veitchianus -a -um, *veitchiorum* see *Veitchia*
velaris -is -e, *velatus -a -um* veiling, veiled
vellereus -a -um densely long-haired, fleecy (*vellus*, a fleece)
velox rapid-growing
Veltheimia for August Ferdinand Graf von Veltheim (1741–1808), German patron of botany
veluti- down-like-
velutinosus -a -um, *velutinus -a -um* with a soft silky down-like covering, velvety
venator of the hunter (the flowers of *Rhododendron venator* are 'hunting-pink')
venenatus -a -um poisonous
venenosus -a -um very poisonous
venetus -a -um of Venice, Venetian
Venidium Veined (the decurrent leaf base on the stem)

venosus -a -um conspicuously veined
Ventenata for Etienne Pierre Ventenat (1757–1808), French botanist
ventilator flabellate, fan-shaped, winnower-like
ventri- belly-
ventricosus -a -um bellied out below, distended to one side, expanded, ventricose
ventriculosus -a -um slightly bellied
venulosus -a -um finely-veined
venustus -a -um graceful, beautiful, charming
Veratrum False-hellebore
verbanensis -is -e from the area of Lake Maggiore (*Lacus Verbanus*)
verbasci- Mullein-like, resembling *Verbascum*
Verbascum a name in Pliny
Verbena the Latin name for the leafy twigs used in wreaths for ritual use and medicine (used by Virgil and Pliny for *Verbena officinalis*)
verbenaca, *verbeni-* from a name in Pliny, vervain-like
verbeniflos *Verbena*-flowered
Verbesina *Verbena*-like (resembles some species)
verecundus -a -um modest
veris of Spring (flowering time), genuine, true, standard
verlotiorum for Verloti, who introduced *Artemesia verlotiorum* from China
vermi- worm-like-, worm-
vermicularis -is -e, *vermiculatus -a -um* worm-like
vernalis -is -e of Spring (flowering time), vernal
verniciferus -era -erum producing varnish
vernicifluus -a -um from which flows a varnish
vernicosus -a -um glossy, varnished
vernix varnish
Vernonia for William Vernon (d. 1711), English botanist
vernonoides *Vernonia*-like
vernus -a -um of the Spring
veronic-, *veronici-* *Veronica*-like
Veronica for St Veronica, who wiped the sweat from Christ's face

verrucosus -a -um with a warty surface, warted, verrucose
versi- several-, changing-
versicolor varying or changeable in colour
verticill-, verticilli- with whorls of-, whorled-
verticillaris -is -e, verticillatus -a -um having whorls (several leaves or flowers all arising at the same level on the stem), verticillate
verticillaster with whorls of flowers
verticillatus -a -um arranged in whorls, verticillate
veruculatus -a -um cylindric and somewhat pointed, like a small pike
veruculosus -a -um somewhat warty
verucund shy, modest
verus -a -um true, genuine
verutus -a -um shaped like a javelin, armed with javelin-like structures
vescus -a -um small, feeble, undernourished; edible
vesicarius -a -um inflated, bladder-like
vesicatorius -a -um blistering (sap causing a local allergic reaction)
vesiculosus -a -um inflated, composed of little blisters
vespertilionis -is -e, vespertilis -is -e bat-like, *vespertilio* (with two large lobes)
vespertinus -a -um of the evening (evening-flowering)
vestae for Vesta, a Roman goddess of the household (Vestal Virgins)
vestalis -is -e white, chaste
vestia for L.C. de Vest (1776–1840), Professor at Graz
vestitus -a -um covered, clothed (with hairs)
Vetiveria Latinized English version of southern Indian name for khus-khus grass
Vetrix Osier
vetulus -a -um old, wizened
vexans annoying, wounding
vexillaris -is -e with a standard (as the large 'sail' petal of a pea-flower)
vialis -is -e, viarum ruderal, of the wayside
viaticus -a -um from the road-side
viatoris is -e of the road-ways, of travellers

viburnoides Viburnum-like

Viburnum the Latin name for the wayfaring tree, *Viburnum lantana*

Vicia the Latin name for a vetch

viciae-, vicii-, vicioides vetch-like-, resembling *Vicia*

vicinus -a -um neighbouring

Victoria, victoriae for Queen Victoria (1819–1901)

Victoriae-Mariae for Queen Mary (Victoria Mary Augusta Louise Olga Pauline Caludine Agnes of Teck), wife of George V

victorialis -is -e victorious (protecting)

Vigna for Dominico Vigna (d. 1647), Professor of Botany at Pisa

vilis -is -e common, of little value

Villarsia for Dominique Villars (1745–1814), Professor at Grenoble, France

villicaulis -is -e with a shaggy stem

villiferus -ero -erum carrying shaggy hairs, clad in a shaggy coat

villipes with a long-haired stalk

villosipes with a very hairy stalk

villosulus -a -um slightly hairy, finely villous

villosus -a -um with long rough hairs, shaggy, villous

vilmorinianus -a -um, vilmorinii for the French nurserymen Vilmorin–Andrieux

viminalis -is -e, vimineus -a -um with long slender shoots suitable for wicker or basketwork, of osiers, osier-like

vinaceus -a -um of the vine, wine-coloured

Vinca Binder (the ancient Latin name refers to its use in wreaths)

Vincetoxicum Poison-beater (the supposed antidotal property of *Vincetoxicum officinale* to snakebite)

vinci-, vincoides periwinkle-like, resembling *Vinca*

vinculans binding, fettering

vindobonensis -is -e from Vienna (*Vindobona*), Viennese

vinealis -is -e of vines and the vinyard, growing in vineyards

vinicolor wine-red

vinifer -era -erum wine-bearing

vinosus -a -um wine-red, wine-like

Viola the Latin name applied to several fragrant plants
violaceus -a -um violet-coloured
violescens turning violet
violoides Viola-like
viorna road ornament, from the French name for traveller's joy, *Clematis vitalba*
viperatus -a -um viper-like (markings)
viperinus -a -um snake's, serpent's
virens, *-virens* green, -vigorous
virescens light-green, turning green
virgatus -a -um with straight slender twigs, twiggy
virgaurea, *virga-aurea* rod-of-gold, golden-rod
virginalis -is -e, *virgineus -a -um* maidenly, purest white, virginal
virginianus -a -um, *virginiensis -is -e* from Virginia, USA, Virginian
virginicus -a -um from the Virgin Islands (West Indies), Virginian
virgulatus -a -um twiggy, striped
virgultorum of thickets
viridescens becoming green, turning green
viridi-, *viridis -is -e*, *viridus -a -um* youthful, fresh-green
viridior more green, greener
viridulus -a -um greenish
virmiculatus -a -um vermillion
virosus -a -um slimy, rank, poisonous, with an unpleasant smell
Viscaria Bird-lime (the sticky stems of German catchfly)
viscatus -a -um clammy
viscidi-, *viscidus -a -um* sticky, clammy, viscid
viscidulus -a -um slightly sticky, somewhat viscid
viscosus -a -um sticky, viscid
Viscum the ancient Latin name for mistletoe or the birdlime from its berries
vistulensis -is -e from the environs of the River Vistula
vitaceus -a -um vine-like, resembling *Vitis*
vitalba vine-of-white (old generic name for the appearance of fruiting *Clematis vitalba*)

Vitaliana, vitalianus -a -um for Vitaliano Donati (1717–1762), Professor at Turin

vitellinus -a -um dull reddish-yellow, egg-yolk-yellow

Vitex an ancient name used in Pliny for Chaste tree, *Vitex agnus-castus*

viti-, vitoides vine-like, resembling *Vitis*

viticella small vine

viticenus -a -um, viticoides *Vitex*-like

Viticola, viticolus -a -um Vine-parasite, inhabiting the vine

viticulosus -a -um sarmentose; producing tendrils

vitiensis -is -e from the Fijian Islands (Viti Levu)

vitigineus -a -um growing as a vine

Vitis the Latin name for the grapevine

vitis-idaea Theophrastus' name for a vine (αμπελος παρα Ιδης, vine from Mt Ida, Greece)

vitreus -a -um, vitricus -a -um glassy

vittae- banded-, filleted-, ribboned-

Vittaria Ribbon

vittarioides *Vittaria*-like

vittatus -a -um striped lengthwise, banded longitudinally

vittiformis -is -e band-like

vittiger -era -erum bearing lengthwise stripes

vivax long-lived (flowering for a long time)

viviparus -a -um producing plantlets (often in place of flowers or as precocious germination on the parent plant), viviparous

vivus -a -um enlivened, long-lasting

vix- hardly-

Voandzeia from the Madagascan name for the underground bean

Vogelia, vogelii for J.R.T. Vogel (1812–1841) of the 1841 Niger expedition

volgaricus -a -um from the river Volga, Russia

volubilis -is -e entwining, enveloping

volutaris -is -e, volutus -a -um with rolled leaves, rolled

vomeformis -is -e, vomiformis -is -e shaped like ploughshares

vomerculus -a -um like a small ploughshare

vomerensis -is -e from Vomero, Naples
vomitorius -a -um causing regurgitation, emetic
Vriesia for Willem H. deVries (1806–1862), Dutch botanist
vulcanicus -a -um fiery, of volcanoes or volcanic soils
vulgaris -is -e, vulgatus -a -um usual, of the crowd, common, vulgar
vulnerarius -a -um, vulnerum of wounds (wound healing property)
vulnerus -a -um marked, wounded (*vulnus* a wound)
vulparia fox-bane (*vulpes* fox)
Vulpia for Johann Samuel Vulpius (1760–1846), German botanist
vulpinus -a -um fox-like, of the fox (colouration, shape of an inflorescence, inferiority)
vulvaria cleft, with two ridges, of the vulva (the smell of *Chenopodium vulvaria*)

Wahlenbergia for Georg Wahlenberg (1780–1851), Swedish botanist
Waldsteinia for Count Franz Adam Waldstein-Wartenberg (1759–1823), Austrian botanist and writer
wallichianus -a -um, wallichii for Nathaniel Wallich (1786–1854), Danish botanist and author
wardii for Frank Kingdon-Ward (1885–1958), collector of E Asian plants, and for Dr Nathaniel Bagshaw Ward (1791–1868), inventor of the Wardian Case
warleyensis -is -e of Warley Place, Essex, home of Miss Ellen Ann Willmott (1858–1934), who developed the garden without regard to financial cost and produced many new cultivars there
Washingtonia for George Washington (1732–1799), American President
watereri for the Waterer Nursery
watermaliensis -is -e from Watermal, Belgium
Watsonia, watsonianus -a -um, watsonium for Sir William Watson (1715–1787), student of sciences
watsonii for William Watson (1858–1925), Curator of Royal Botanic Gardens, Kew

weddellianus -a -um, weddellii for Dr Hugh Algernon Weddell (1819–1877), botanist and traveller

Wedelia for George Wolfgang Wedel (1645–1721), Professor at Jena

Weigela for Christian Ehrenfried von Weigel (1748–1831), Professor at Griefswald

whipplei for Lieut. Amiel Weeks Whipple (1818–1863), pioneer on the transcontinental American railway

wichuraianus -a -um for Max E. Wichura (1817–1866), German botanist

Widdringtonia for Capt. Widdrington, botanist explorer

Williamsii for a number of plant collectors, plant introducers, breeders and gardeners of whom Benjamin Samuel Williams (1824–1890) was an orchidologist (author of *The Orchid-grower's Manual*), Percival Dacre Williams (1865–1935) created the garden at Lanarth in Cornwall and his cousin J.C. Williams of Caerhays Castle

willmottiae, willmottianus -a -um for Miss Ellen Anne Willmott (1858–1934), rose specialist and plant introducer

wilsonii for several including Dr Ernest Henry Wilson (1876–1931) who collected in the east for Veitch and the Arnold Arboretum and George Fergusson Wilson (1822–1902), founder of the Wisley Wild Garden

wilsoniae for Mrs Ernest H. Wilson

Winteri for Capt. Winter who sailed in Drake's ship (*Drimys winteri*)

wintonensis -is -e from Winchester, *Venta*

Wisteria for Caspar Wister (1761–1818), anatomist of Pennsylvania University

wittrockianus -a -um for Prof. Veit Brecher Wittrock (1839–1914), of Stockholm, Director of Hortus Bergianus and writer on pansies

Wolffia for J.F. Wolff (1778–1806), German doctor, or Herman Wolff (1866–1929), veterinary surgeon and botanist

wolfianus -a -um, wolffii for Ferdinand Otto Wolf (1838–1906), Professor of Botany at Sitten; and for Franz Theodor Wolf (b. 1841), German geologist and botanist

wolgaricus -a -um from the region of the River Volga, Russia
Woodsia for Joseph Woods (1776–1864), English botanist
Woodwardia for Thomas Jenkinson Woodward of Suffolk (1745–1820), British botanist
Wulfenia for Franz Xavier Freiherr von Wulfen (1728–1805), Austrian botanical writer

xalapensis -is -e from Xalapa, Mexico; see *jalapa*
xanth-, xanthi-, xantho- yellow- (ξανθος)
xanthacanthus -a -um yellow-thorned
Xantheranthemum Yellow-*Eranthemum* (or Lovely-yellow-flower)
xanthinus -a -um yellow
Xanthium Dioscorides' name, ξανθιον, for cocklebur, from which a yellow hair dye was made
Xanthoceras Yellow-horn (the processes on the disc)
Xanthorhiza Yellow-root
xanthospilus -a -um yellow-spotted
xanthostephanus -a -um with a yellow crown
xanthoxylum (on) yellow-wooded
xeno- foreign-, unnatural-, strange- (ξενος)
xer-, xero- dry- (ξηρος)
Xeranthemum Dry-flower (Immortelle)
xerophilus -a -um drought-loving, living in dry places
Xerophyllum Dry-leaf (*Xerophyllum* × *tenax* Elk grass)
xestophyllus -a -um having polished leaves
xiphioides *Xiphium*-like
xiphium (on) Sword, ξιφος (old generic name from the Greek name for a *Gladiolus*)
xiphochilus -a -um with a sword-shaped lip
xiphoides sword-like, shaped like a sword
Xiphopteris Sword-fern
xiphophyllus -a -um with sword-shaped leaves
Xolisma an uncertain name for a genus containing species with such vernacular names as 'maleberry', 'fetterbush' and 'staggerbush'

xylo-, -xylon, -xylum wood-, ξυλον, woody-, -wooded, -timbered
xylocanthus -a -um woody-thorned
xylophilus -a -um wood-loving (of wood-attacking fungi)
Xylopia Bitter-wood (the Greek for such wood)
xylopicron bitter-wooded, ζυλοπιχρος
xylosteum hard-wooded (ξυλοςτεον wood-bone)
Xyris Greek name used by Dioscorides for *Iris foetidissima*
Xysmalobium Fragmented-lobes (of the corona)

yakusimanus -a -um from Yakushima, Japan
yedoensis -is -e from Tokyo (Yedo), Japan
yemensis -is -e from the Yemen, Arabia
yosemitensis -is -e from the Yosemite Valley, California, USA
youngianus -a -um for the Youngs, nurserymen on the Milford estate, Epsom until 1862
Yucca from a Carib name for cassava (*Manihot*), for its enlarged roots
yuccifolius -a -um with *Yucca*-like leaves
yulan from the Chinese name for *Magnolia denudata*
yunnanensis -is -e from Yunnan, China

za- much-, many-, very-
zaleucus -a -um very white
zalil from an Afghan vernacular name for *Delphinium zalil*
Zaluzianskya for Adam Zalusiansky von Zalusian (1558–1613), of Prague, author of *Methodus Herbariae*
zaman as *saman*, from a South American vernacular name
zambac as *sambac*, from an Arabic vernacular name
Zamia Loss, a name in Pliny refers to the sterile appearance of the staminiferous cones
zamii- resembling *Zamia*
Zannichellia for Zannichelli, Italian botanist
Zantedeschia for Giovani Zantedeschi (1773–1846), Italian doctor and botanist
zantho- yellow (ξανθος)

Zanthorhiza (Zanthorrhiza) Yellow-root
Zanthoxylum Yellow-wood
zanzibarensis -is -e, zanzibaricus -a -um from Zanzibar, East Africa
zapota from the Mexican name, cochil-zapotl, for the chicle tree, *Achras sapota*
Zauschneria for Johann Baptist Joseph Zauschner (1737–1799), Professor at Prague
zazil (zalil) from an Afghan name for a *Delphinium*
Zea from the Greek name for another grass, possibly spelt
zebrinus -a -um from the Portuguese, meaning striped with different colours, zebra-striped
Zelkova from the Caucasian name, tselkwa, for *Zelkova carpinifolia*
Zenobia an ancient Greek name, Zenobia was Queen of Palmyra
Zephyranthes West-wind-flower (ζεφυρος)
zephyrius -a -um western, flowering or fruiting during the monsoon season (for Indonesian plants)
Zerna a Greek name, ζερνα (for the *Cyperus*-like spikelets)
zetlandicus -a -um from the Shetland Isles
zeylanicus -a -um from Sri Lanka (Ceylon), Singhalese
zibethinus -a -um of the civet (the foul-smelling fruits of *Durio zibethina* are used to trap the Asiatic civet (*Vivera zibetha*)
Zigadenus Yoked-glands (paired glands at the perianth base)
zigomeris -is -e see *zygomeris*
zimapani from Zimapan, Mexico
Zingiber from a pre-Greek name, possibly from Indian or oriental source, inchi (a root)
Zinnia for Johann Gottfried Zinn (1727–1759), Professor of Botany at Göttingen
Zizania an ancient Greek name, ζιζανιον, for a wild plant
zizanoides resembling *Zizania* (Canadian wild rice)
Zizyphus ancient Greek name, ζιζυφος, for *Ziziphus jujuba* (the Arabic for *Z. lotus* is zizouf)
zoysii for Karl von Zoys
zonalis -is -e girdled with distinct bands or concentric zones

zonatus -a -um with zones of colour markings
zooctonus -a -um poisonous, ζωο-κτονος, life killing
zoster- girdle-
Zostera Ribbon (Theophrastus' name, ζωστηρ, for a marine plant)
zygis yoke-like (paired flowers)
Zygnema Paired-thread (at conjugation)
zygo- paired-, balanced-, yoked- (ζυγος)
Zygocactus Jointed-stem-cactus
zygomeris -is -e with twinned parts
zygomorphus -a -um bilateral, of balanced form
Zygophyllum Yoked-leaves (some have conspicuously paired leaves)

Addendum to glossary

aberans differing, deviating from the norm
acalycinus -a -um lacking a calyx
achypodus -a -um scaly-stemmed, chaffy-stemmed, αχυ-ποδος chaff
acpunctus -a -um spotted above, with spots towards the apex
acraeus -a -um useless, weak, αχρειος
adroseus -a -um near *roseus -a -um*
aequinoctiianthus -a -um flowering about the time of the equinox
agapetus -a -um desirable, love, αγαπη
agnatus -a -um related, offspring of the father
agrippinus -a -um for Marcus Vispanius Agrippa (63–12 BC), Roman General
akakus -a -um harmless, innocent, ακακος
alienus -a -um of others, strange, alien
allantophyllus -a -um with sausage, αλλας, or variable, αλλαγ-, -shaped leaves
alpigenus -a -um born of the mountains
altis -is -e above, on high, from afar
alticolus -a -um inhabiting high places
amarissimus -a -um most bitter
amblyanthus -a -um feeble flowering, blunt-flowered, αμβλυ-αντηος
ananassus -a -um pineapple-like (the fruiting receptacle)
anastreptus -a -um twisted backwards, curved upwards, ανα-στρεπτος
anatinus -a -um healthy, rewarding, ανα-τινω
anchoriferus -a -um with flanges, bearing anchors
anguifugus -a -um snake-refuge
anthiodorus -a -um fragrant-flowered

anticarius -a -um   from the area of Caria
aperantus -a -um   open-flowered
aratophyllus -a -um   with plough-like leaves, *arantor* a ploughman
arcanus -a -um   of coffins, of boxes, of cages (use in basketry)
aretioides   resembling *Aretia (Androsace)*
argolicus -a -um   from the area of Argos, S Greece
argun   sharp
arifolius -a -um   *Arum*-leaved
arpadianus -a -um   sickle-like, αρπη (leaves)
aspalathoides   like a thorny shrub, ασπαλαθος
assoanus -a -um   from Aswan, Egypt
atomarus -a -um   unmoved, indivisible
auct.   used by a writer to indicate that the name he cites is used by other authors not in the sense he supports
avellanidens   with tearing teeth
azedarach   from a Himalayan vernacular name for the bead tree

barystachys   heavily-branched
basjoo   Japanese name for the fibre of *Musa basjoo*
bathyphyllus -a -um   thick-leaved, βαθυ-φυλλον
biseptus -a -um   cleft in two, deeply divided into two
boldus -a -um   from Araucarian vernacular name, voldo, for the fruit of *Peumus boldo*
bombicis -is -e   of silk (silk-worms feed on *Morus bombicis*)

diaprepes   distinguished, conspicuous, excellent, διαπρεπω
dichlamydius -a -um   two-cloaked, δι-χλαμυδος (double spathe)
dilatatopetiolaris -is -e   with inflated petioles
dimitrus -a -um   two-turbanned
doliiformis -is -e   tubby, wine-jar-shaped (*dolium*)
dracocephalus -a -um   dragon-headed
dracophyllus -a -um   dragon-leaved (markings)
drumonius -a -um   of woodlands, δρυμος
duratus -a -um   hardy

egregius -a -um outstanding, exciting
eldorado golden-one
encliandrus -a -um enclosed-flower
enneaphyllus -a -um nine-leaved
ensete from the Abyssinian vernacular name for *Musa ensete*
epipremnus -a -um of tree-stumps, growing on tree-trunks, επι-πρεμνον
episcopalis -is -e worthy of attention, of bishops
euchrous -a -um well-coloured
eucrosioides resembling *Eucrosia* (Amaryllidaceae)
eutriphyllus -a -um three-leaved throughout
evenosus -a -um without conspicuous veins
explanatus -a -um distinct
exserens thrusting out, revealing
exsul foreigner, exile

fabri skilfully produced, for building (timber of *Abies fabri* and *Acer fabri*)
falcinellus -a -um little-scythe-like (pinnae)
fascinator magic, very interesting, bundled
faucius -a -um throated (corolla)
figo fixed, pierced
firmipes strong-stemmed, stout-stemmed
flaccospermus -a -um with *flacca*-like seeds (*Carex*)
flandrius -a -um from Flanders, France, Flandrian
flavonutans yellow-drooping
forniculatus -a -um slightly arched
frutetorus -a -um of scrubland
funiferus -a -um cord-bearing
fusculus -a -um husky, blackish

galeritus -a -um of rustic places
geminiflorus -a -um with paired flowers
gibbiflorus -a -um convex-flowered (reflexed petals)

ginseng from the Chinese name for the drug
glebarius -a -um lump, clod, sod, soil
gnomus -a -um dwarfish
granatus -a -um pale-scarlet
gryllus -a -um of the cricket, (haunt or scabrid surface)
guttulatus -a -um slightly glandular
gypsicolus -a -um gypsum-loving, calcium-loving, γυπσος

habrotrichus -a -um softly-hairy, soft-haired
hamabo from Hama, N Syria
hamadae together, neighbours, binding (the uses of rattan)
hamosus -a -um hooked
haspan from a Ceylonese vernacular name
hectori, hectorii, hectoris -is -e for Hector, son of Priam of Troy
heptapeta a misnomer by Buc'hoz to describe *Lassonia heptepeta* from a picture showing only five erect tepals (petals)
hesperidus -a -um *Citrus*-fruited (of the guardians of the golden apple tree, of Spain)
homonymus -a -um of the same name, ομο-ονομα, ομονυμος
horridulus -a -um somewhat thorny, prickly
hymenaeus -a -um for the God of marriage, of weddings

idomenaeus -a -um for Idomeneus, King of Crete
ignavus -a -um igniting (autumn tints)
insanus -a -um outrageous, frantic
insolitus -a -um unusual (*solitus* normal)
intaminatus -a -um chaste, unsullied
iocastus -a -um for Jocasta, mother and wife of Oedipus
iodes violet-coloured (ιον a violet)
irroratus -a -um dewy (*irroro* to bedew)
ischnopus -a -um thin-stalked, with slender stems

jaburan from the Japanese name for *Ophiopogon* jaburan
jujuyensis -is -e from Jujuy province of NW Argentina

kapelus -a -um of the retailer or innkeeper, καπηλος
karroo from the name, Karoo, of the S African semi-desert plateau
karooensis -is -e, karooicus -a -um from the S African Karoo
kurroo from the Gahrwain vernacular for *Gentiana kurroo*

lagenicaulis -is -e with flask-shaped stems
lahue from the Lahu region of SW China
lapidicolus -a -um living in stony places
lapidosus -a -um of stony places
leiomerus -a -um with smooth parts, smooth
leptolepis -is -e with slender scales
lobularis -is -e with small lobes
lotifolius -a -um *Lotus*-leaved
luna of the moon, crescent-shaped, of a month
lysolepis -is -e with loose scales, λις, λισσος

macrorrhabdos large-stemmed, heavily branched
magenteus -a -um from Magenta, N Italy
majusculus -a -um somewhat larger
malortianus -a -um of the preferred
manniferus -a -um manna-bearing, μαννα
masoniorus -a -um of the Masons
megathurus -a -um large-tailed
mirus -a -um wonderful, strange
miserus -era -erum wretched, poor, pitiful
moschus -a -um fragranced, musk-like
myrobalanus -a -um perfumed-nut, perfumed acorn, μυρο-βαλανος

naso a name used by Ovid
nativo created, native, natural
neurolobus -a -um with veined lobes
nigercors with a black centre, black-hearted

nigramargus -a -um with black edges
nitrariaceus -a -um of alkaline soils

oedogonatus -a -um with swollen nodes, οιδημα γονιμος (compares with the swollen reproductive organs on filamentous alga *Oedogonium*)
oistophyllus -a -um with sagittate leaves, with arrow-shaped leaves, οιστο-θυλλον
oligodon few-toothed, feebly-toothed
onychinus -a -um onyx-like (layers of colour)
ophianthus -a -um long-flowered, serpentine-flowered, οφι-ανθος
opiparus -a -um sumptuous
orbus -a -um orphaned, of grooves
ornans embellished, like manna-ash
orphanidis -is -e destitute, of the bereaved, ορφανος
oscularis -is -e kissing (the touching leaf margins)
otocarpus -a -um with ear-like fruits, with shell-like fruits
otolepis -is -e with ear-shaped scales, with shell-like scales

pagophilus -a -um rural, village-loving
palibinianus -a -um from the Hawaiian for cliffs, of cliffs
panteumorphus -a -um well formed all round
parilicus -a -um equal
pavius -a -um from Pavia, Italy
pavimentatus -a -um pavement, paving (mode of growth)
peculiaris -is -e special
penna-marina sea-feather
perado from a vernacular name on the Canary Isles
perdulce very sweet, pleasant throughout
peregrinans wandering, spreading
peuce an ancient Greek name for *Pinus*
phaedropus -a -um of gay appearance, jolly-looking, φαιδροωπης
phlogopappus -a -um bearing red down

pholidotus -a -um with scaly ears (bracts)
phycofolius -a -um seaweed-leaved, φυκος + *folius*
phyctidocalyx with a deciduous calyx, φυκτοδο-καλυξ
pilosulus -a -um somewhat hairy, loosely pilose
pinguifolius -a -um waxy-leaved, thick-leaved
pisacomensis -is -e from the Pisa area, Italy
planetus -a -um flat, with fully opening flowers
pleistranthus -a -um long-flowering, heavily-flowering, πλειστρο-ανθος
plumeus -a -um feathery
pocophorus -a -um woolly, fleece-bearing, ποκο-φορα
polyepsis -is -e long-flowering, of many dawns, πολυ-εψος
praeflorens early-flowering
primavernus -a -um the first of Spring
primiveris -is -e the first of Spring
protistus -a -um number one, the first of the first, πρωτος, προτιστος
pterostoechas winged-*stoechas*
pugniformis -is -e fist-shaped (*pugnus* a fist)
pumilio very small
punctorius -a -um dotted, spotted, punctate
pyrami for Pyramus, Thisbe's lover

quadrifarius -a -um four-partite, four-ranked
quilus -a -um tube-like
quinoa Andean vernacular name, kanua, for *Chenopodium quinoa*
quinquepeta a misnomer by Buc'hoz who misinterpreted a picture as having only seven tepals. see *heptepeta*

rajah from the Hindi vernacular word for King
ramellosus -a -um like brushwood, twiggy
redolens promising, perfumed
rhombeus -a -um shaped like a rhombus, ρομβος
rifanus -a -um of the Rif, N African tribe

rigidulus -a -um quite stiff
rotang Indian vernacular name for rattan vine
rubis -is -e reddish

sancti-johannis Saint John's (wort)
scotostictus -a -um with dark spots, dark-punctured, σκοτο-στικτος
selensis -is -e moonshine
sentis -is -e briar-like, thorny
sericeovillosus -a -um silky-hairy
sexflorus -a -um six-flower, with six-flowered racemes
sicyoides from Sicyon, Peloponnese
sierrae of the Sierras, grassland
simorrhinus -a -um monkey-snouted (spathe)
simulatrix imitator (*Salix* being feminine)
sopherus -a -um Arabian name for a pea-flowered tree
sorediatus -a -um heaped, mounded (as lichen soridea)
sphacioticus -a -um damaged-looking, gangrened, σφακελος
spiculifolius -a -um with spicule-like leaves
spilotus -a -um stained-looking, σπιλος
squamosorodicosus -a -um with gnawed scales, irregular scales
staphisagrius -a -um wild-grape-like, σταφυς-αγριος
stauropetalus -a -um cruciform-petalled, petals forming a cross
stenaulus -a -um narrow-tubed, στεν-αυλον
stimulosus -a -um goad-like, pole, like
studiosorus -a -um of specialists
sublatus -a -um lofty
succirubrus -a -um with reddish-amber sap (*succus*)
sumulus -a -um excellent
surrepans sprawling, not quite creeping
syrticolus -a -um of sand-banks

talasicus -a -um winged, enduring, wretched
tananicus -a -um from the River Don, *Tanais*, Sarmatia

tapetodes carpeting
telopeus -a -um conspicuous, seen from afar
tenius -a -um persisting, tenacious
tetrastichus -a -um with four rows, four-ranked
thaumus -a -um marvellous, θαυμα to marvel
thea from the Chinese name, T'e
theezans tea-like, resembling *Thea*
thusculus -a -um from Tusculum, incense
tibae of flutes, from Tuvoli *Tibur*
ticus -a -um fecund, fruitful
tmoleus -a -um from the Tmolus Mountian, Lydia
tolminsis -is -e from Tolmin, former Yugoslavia
toona a former generic name
topiarius -a -um of ornamental gardens
tormentillus -a -um an ancient Latin name, anguish, torment
torus -a -um mounded, knotted, ornamental
totarus -a -um from the New Zealand Maori vernacular name, totara
tranquillans calming
transens latticed, intertwined
transitorius -a -um between, intermediate, transitory
triornithophorus -a -um bearing three birds, the flower-heads
tristaniicarpus -a -um with *Tristania*-like fruits
trojanus -a -um from Troy, Trojan
truliifolius -a -um with leaves shaped like trowels or pans
truncorus -a -um growing on tree trunks, of tree stumps
tuguriorus -a -um of hut-dwellers
tumidissinodus -a -um with very tumid nodes
turbith untidy, crowded, τυρβη
tutelatus -a -um protected

ulophyllus -a -um shaggy-leaved
unguis-cati cat's-clawed, with recurved thorns
uratus -a -um tailed, ουρα

venustulus -a -um quite charming
vicarius -a -um proxy, substitute
vigilis -is -e awake, watching
visnagus -a -um withering, old Norse, visna
vivax inflammable, long-lived
vulcanicolus -a -um living on volcanic soils

Bibliography

Adanson, M. 1763–64 *Familles des Plantes*. Paris.

Bagust, H. 2001 *Plant Names: Common & Botanical*. Helicon, Oxford.

Bailey, L.H. 1949 *Manual of Cultivated Plants*. Macmillan, New York.

Bateson, W. 1909 *Mendel's Principles of Heredity*. Cambridge.

Bauhin, C. 1620 *Prodromus theatri botanici*. Frankfurt.

Bauhin, C. 1623 *Pinax theatri botanici*. Basel.

Boerhaave, H. 1710 & 1720 *Index plantarum* .... Leiden.

Brickell, C.D. et al. 1980 International Code of Nomenclature for Cultivated Plants. In *Regnum Vegetabile*, Vol. 104. Deventer.

Britten, J. & Holland, R. 1886 *A Dictionary of English Plant Names*. The English Dialect Society, London.

Brummitt, R.K. 1992 *Vascular Plant Families and Genera*. Royal Botanic Gardens, Kew.

Brunfels, O. 1530–36 *Herbarium vivae eicones* .... Strasbourg.

Caesalpino, A. 1583 *De plantis libri* xvi. Florence.

Camp, W.H., Rickett, H.W. & Weatherby, C.A. (eds.) 1947 Rochester Code. *Brittonia* **6**(1). 1–120. Chronica Botanica, Mass., USA

Candolle, A. de 1867 *Lois de la nomenclature botanique*. H. Georg, Paris.

Candolle, A.P. de 1813 *Théorie élémentaire de la botanique*. Paris.

Chittenden, F.J. (ed.) 1951 *Royal Horticultural Society Dictionary of Gardening*, Vols. 1–4, and Supplements 1956 and (ed. P.M. Synge) 1969. Oxford University Press.

Cordus, V. 1561–63 *Annotationes in pedacii Dioscorides* .... Strasbourg.

Correns, C. 1900 G. Mendel's Regel über das Verhalten der Nachkommenschaft der Rassenbastarde. *Berichte* 18, 158.

Cube, J. von 1485 *German Herbarius*. Mainz.

Darwin, C.R. 1859 *The origin of species by means of natural selection*. J. Murray, London.

Dioscorides, P. *De Materia Medica*, John Goodyer translation of 1655 (ed. R.T. Gunther). Oxford University Press, Oxford.

Dodoens, R. 1583 *Stirpium historiae pemptades*. Antwerp.

l'Ecluse, C. 1583 *Stirpium nomenclator pannonicus*. Német-Hjvár.

Farr, E.R. et al. 1979–86 Index Nominum Genericorum and Supplement 1. In *Regnum Vegetabile*, Vols. 100, 101, 102, and 113. The Hague.

Fernald, M.L. 1950 *Gray's Manual of Botany*. The American Book Company, New York.

Fuchs, L. 1542 *De historia stirpium* . . . . Basel.

Gilbert-Carter, H. 1964 *Glossary of the British Flora*, 3rd edn. Cambridge University Press, Cambridge.

Greatwood, J., Hunt, P.F., Cribb, P.J. & Stewart, J. (eds.)1985 *The Handbook on Orchid Nomenclature and Registration*, 3rd edn. The International Orchid Commission, London.

Green, M.L. 1927 The history of plant nomenclature. *Kew Bulletin*, 403–415.

Greuter, W. et al. *2000 International Code of Botanical Nomenclature (St Louis Code).* Koeltz Scientific Books, Königstein, Germany.

Grew, N. 1682 *The Anatomy of Plants*. London.

Grigson, G. 1975 *An Englishman's Flora*. Hart Davis, St Albans.

Ivimey-Cook, R.B. 1974 *Succulents -- A Glossary of Terms and Descriptions*. The National Cactus and Succulent Society, Oxford.

Jackson, B.D. 1960 *A Glossary of Botanical Terms*, 4th edn. Duckworth, London.

Jeffrey, C. 1978 *Biological Nomenclature*, 2nd edn. Edward Arnold, London.

Johnson, A.T. & Smith, H.A. 1958 *Plant Names Simplified*. Feltham.

Jung, J. Doxoscopiae; 1662; Isagoge phytoscopica 1679. In *Opuscula botanica-physica*. Coburg 1747

Jussieu, A.L. 1789 *de Genera Plantarum*. Paris.

Linnaeus, C. 1735 Systema naturae. Leiden.

Linnaeus, C. 1738 Classes plantarum. Leiden.

Linnaeus, C. 1751 Philosophia botanica. Stockholm/Amsterdam.

Linnaeus, C. 1753 Species plantarum. Stockholm.

Linnaeus, C. 1754 Genera plantarum. edn.5 Stockholm.

Linnaeus, C. 1759 Systema naturae. edn.10 Stockholm.

Linnaeus, C. 1762–3 Species plantarum. edn.2 Stockholm.

Linnaeus, C. 1764 Genera plantarum. edn.6. Stockholm.

MacLeod, R.D. 1952 *Key to the Names of British Plants*. Pitman, London.

Magnus, Albertus 1478 *Liber aggregationis seu liber secretorum Alberti magni de virtutibus herbarum*. Johann de Anunciata de Augusta.

Malpighi, M. 1671 . . . anatome plantarum . . . . London (1675–79).

Mendel, G.J. 1866 *Versuche über Planzenhybriden*. Brno.

Mentzel, C.M. 1682 *Index Nominum Plantarum Multilinguis (Universalis)*. Berlin.

Morison, R. 1672 *Plantarum Umbelliferum distributio nova* . . . Oxford.

Morison, R. 1680 *Plantarum historia Universalis* . . . . Oxford.

l'Obel, M. 1576 *Plantarum seu stirpium historia* . . . . Antwerp.

Pankhurst, A. 1992 *Who does your Garden Grow?* Earl's Eye Publishing, Colchester.

Paracelsus (Bombast von Hohenheim) 1570 *Dispensatory and Chirurgery . . . Faithfully Englished* by W.D. London 1656.

Parkinson, J. 1629 *Paradisi in Sole Paradisus Terrestris*. Lownes H. & Young R. London (reprinted by Methuen, London, 1904).

Gaius Pliny Secundus (AD 23–79) Thirty seven books of *Historia Naturalis*, of which 16 deal with plants.

Plowden, C.C. 1970 *A Manual of Plant Names*. George Allen & Unwin, London.

Porta, Giambatista della (Johannes Baptista) 1588 *Phytognomica*. Naples.

Prior, R.C.A. 1879 *On the Popular Names of British Plants*, 3rd edn. London.

Rauh, W. 1979 *Bromeliads*. English translation by P. Temple. Blandford Press, Dorset.

Ray, J. 1682 *Methodus plantarum*. London.

Ray, J. 1686–1704 *Historia plantarum*. London.

Rivinus, A.Q. 1690 *Introductio generalis in rem herbariam*. Leipzig.

Schultes, R.E. & Pease, A.D. 1963 *Generic Names of Orchids--Their Origin and Meaning*. Academic Press, London.

Smith, A.W. 1972 *A Gardener's Dictionary of Plant Names* (revised and enlarged by W.T. Stearn). Cassell, London.

Sprague, T.A. 1950 The Evolution of Botanical Taxonomy from Theophrastus to Linnaeus. In *Lectures on the Development of Taxonomy*. Linnean Society of London.

Stafleu, F.A. et al. (eds.) 1983 International Code of Botanical Nomenclature. In *Regnum Vegetabile*, Vol. 111. Utrecht.

Stafleu, F.A. & Cowan, R.S. 1976– Taxonomic Literature. edn.2. In *Regnum Vegetabile* Vol.94 (1976),Vol.98 (1979),Vol.105 (1981),Vol.110 (1983),Vol.112 (1985),Vol.115 (1986) . . . Utrecht.

Stearn, W.T. 1983 *Botanical Latin*. David & Charles, Newton Abbot.

Stearn, W.T. 1992 *Stearn's Dictionary of Plant Names for Gardeners*. Cassell, London.

Styles, B.T. (ed.) 1986 *Infraspecific Classification of Wild and Cultivated Plants*. The Systematics Association, Special Vol. 29. Oxford.

Sutton, W.S. 1902 On the morphology of the chromosome group in *Brachystola magna*. *Bio.Bull.* 4, 24–39.

Theophrastus 1483 *De causis plantarum*, lib VI. Bartholomaeum Confalonerium de Salodio.

Tournefort, J.P. de 1694 *Elemens de botanique*. Paris.

Tournefort, J.P. de 1700 *Institutiones rei herbariae*. Paris.

Trehane, R.P. 1995 *International Code of Nomenclature for Cultivated Plants*. Quarterjack Publishing, Wimborne, Dorset.

Tschermak, E. 1900 Uber Künstliche Kreuzung bei *Pisum sativum*. *Biologisches Zentralblatt*, 20, 593–595.

Turner, W. 1538 *Libellus de Re Herbaria* and

Turner, W. 1548 *The Names of Herbes* (The Ray Society, Vol. 145, London. 1965).

Turner, W. 1551–68 *A New Herbal*. London & Cologne.

U.P.O.V. 1985 International convention for the protection of new varieties of plants. Texts of 1961, 1972, 1978. U.P.O.V. publication 293E. Geneva.

deVries, H. 1900 Sur la loi de disjonction des hybrides. *Comptes Rendues*. 130;845–847. Paris. (Das Spaltungsgesetz der Bastarde; Vorlaufige Mitteilung. *Ber. Dtsch. Bot. Ges.* 18; 83–90. 1900.)

Willis, J.C. 1955 *A Dictionary of Flowering Plants and Ferns*, 6th edn. Cambridge University Press, Cambridge.

Wilmott, A.J. 1950 Systematic Botany from Linnaeus to Darwin. In *Lectures on the development of taxonomy*. Linnean Society of London.

Zimmer, G.F. 1949 *A Popular Dictionary of Botanical Names and Terms*. Routledge & Kegan Paul, London.